Lecture Notes in Nanoscale Science and Technology

Volume 45

Lecture Notes in Nanoscale Science and Technology (LNNST) aims to report latest developments in nanoscale science and technology research and teaching – quickly, informally and at a high level. Through publication, LNNST commits to serve the open communication of scientific and technological advances in the creation and use of objects at the nanometer scale, crossing the boundaries of physics, materials science, biology, chemistry, and engineering. Certainly, while historically the mysteries in each of the sciences have been very different, they have all required a relentless step-by-step pursuit to uncover the answer to a challenging scientific question, but recently many of the answers have brought questions that lie at the boundaries between the life sciences and the physical sciences and between what is fundamental and what is application. This is no accident since recent research in the physical and life sciences have each independently cut a path to the edge of their disciplines. As both paths intersect one may ask if transport of material in a cell is biology or is it physics? This intersection of curiosity makes us realize that nanoscience and technology crosses many if not all disciplines. It is this market that the proposed series of lecture notes targets.

Radouane En-nadir
Editor

Quantum Systems at the Nanoscale

Fundamentals, Characterization, Properties, and Emerging Applications

Springer

Editor
Radouane En-nadir
Sherbrooke University
Sherbrooke, QC, Canada

ISSN 2195-2159 ISSN 2195-2167 (electronic)
Lecture Notes in Nanoscale Science and Technology
ISBN 978-3-032-06644-2 ISBN 978-3-032-06645-9 (eBook)
https://doi.org/10.1007/978-3-032-06645-9

This Springer imprint is published by the registered company Springer Nature Switzerland AG
The registered company address is: Gewerbestrasse 11, 6330 Cham, Switzerland

To the scientists, students, and engineers who are building tomorrow's technologies one atom at a time—and to all curious minds who believe that the invisible world holds the key to a brighter future.

Preface

The quest for faster, smaller, and more efficient devices has reshaped the landscape of modern technology. At the heart of this transformation lies a class of materials whose behavior defies classical intuition—materials structured at the nanoscale, where the motion of charge carriers is confined, controlled, and engineered with extraordinary precision.

Compound semiconductors from the III–V and III–N families have long stood out for their exceptional optoelectronic properties. When shaped into low-dimensional systems—quantum wells, quantum dots, and quantum wires—their potential becomes even more remarkable. In these nanostructures, carriers are not free to roam as in bulk materials; they are confined to dimensions where quantum effects dominate. As a result, light–matter interaction can be amplified, losses minimized, and energy levels precisely tailored to match the needs of cutting-edge devices.

Yet, real-world optoelectronic systems—lasers, solar cells, photodetectors—face fundamental limitations: non-radiative recombination, thermal losses, restricted absorption bandwidths, and environmental noise. The promise of low-dimensional systems is not just theoretical—it's practical and urgent. By understanding and controlling carrier dynamics at the nanoscale, we gain new tools to enhance radiative efficiency, extend device lifetime, boost light conversion, and reduce parasitic effects. Devices become not only smaller but smarter, more responsive, and more energy-efficient.

This book brings together a collection of focused studies exploring how carrier dynamics—such as absorption, recombination, and confinement—are influenced by quantum structure geometry, material composition, and external conditions. Across all contributions, a unifying theme emerges: quantum engineering at the nanoscale is central to unlocking the next generation of high-performance photonic and energy devices.

The chapters are structured to progressively build insight, each offering a unique perspective on this rapidly advancing field:

- Chapter "Fascinating New Generation Core–Shell Quantum Dots"—Core–Shell Quantum Dots and Their Applications

Core–shell nanostructures provide enhanced optical stability and quantum yield, positioning them as key candidates for photonic, optoelectronic, and biomedical innovations.

- Chapter "Donor Interaction Physics of Semiconductor QD Nano-Spheres"—Donor Impurity States in Quantum Dots

Impurity charge states significantly influence energy levels and optical transitions, offering design strategies for doped nanostructures in next-generation quantum electronics.

- Chapter "Normalized Exciton Binding Energy in Cylindrical Nanostructures: Influences of Quantum Confinement and LO Phonon Contributions"—Exciton Binding in Cylindrical Nanostructures

Quantum confinement and phonon interactions shape exciton behavior, with far-reaching consequences for efficient light–matter coupling in advanced optoelectronic systems.

- Chapter "The Nonlinear Optical Rectification Coefficient in a MSQD: Effects of External Perturbation and the Kratzer Confinement Potential"—Nonlinear Optical Rectification in Multilayer Quantum Dots

Structural and environmental factors modulate nonlinear optical responses, enriching our understanding of signal processing and light manipulation at the nanoscale.

- Chapter "Hydrostatic Pressure and Temperature Effects on Nonlinear Optical Rectification in Tetrapod Quantum Dots"—Optical Behavior of Core–Shell Tetrapod Nanostructures

Donor doping, mechanical pressure, and temperature shifts influence subband transitions and optical performance in these uniquely shaped nanostructures.

- Chapter "Absorption Coefficient of a Delta-Doped Layer Inside Three Quantum Wells Subjected to an Electric Field"—Coupled Quantum Wells Under Applied Fields

Electronic transitions in δ-doped quantum wells under external electric fields reveal new pathways for tunable optoelectronic components such as solar cells and electro-optic modulators.

- Chapter "Stark-Shift, Binding Energy, Diamagnetic Response, Polarizability and Dipole Moment of Shallow Donor-Impurity in Core/Shell Quantum Disk"—Donor Properties in Core–Shell Quantum Disks

Shell thickness and lateral electric fields shape Stark shifts and polarizability, unveiling new parameters for tuning quantum properties in disk-like nanostructures.

- Chapter "Performance of III-N-Based Quantum Well Intermediate Band Solar Cells"—Intermediate Band Solar Cells Based on Quantum Wells

A comprehensive analysis of physical and technological aspects impacting the performance of intermediate band solar cells (IBSCs), aiming to enhance solar energy conversion.

- Chapter "Recombination Dynamics and Radiative Lifetimes in III–N Semiconductor Photonic Structures"—Recombination and Radiative Lifetime Dynamics in Quantum Wells

Quantum well geometry, alloy composition, and external fields impact radiative lifetimes and transition energies, contributing to performance improvements in InGaN-based devices.

- Chapter "Optical Absorption and Oscillator Strength Modulation in GaAs-Based Quantum Dots"—Kratzer Potential's Impact on Optical Absorption in Multi-quantum Dots
- Investigates the influence of Kratzer confinement potential on the optical absorption properties in multilayer quantum dot systems, highlighting its role in tuning electronic transitions and spectral behavior.
- Chapter "Kratzer Potential's Impact on Optical Absorption in Multilayer Quantum Dots"—Optical Absorption and Oscillator Strength Modulation in GaAs-Based Quantum Dots
- Examines how structural parameters affect optical absorption and oscillator strength in GaAs quantum dots, offering insights into the control of light–matter interaction in semiconductor nanostructures.

Together, these chapters form a coherent narrative—one that bridges fundamental physics with real-world applications. We hope that this book will serve as both a reference and a source of inspiration for those aiming to harness the extraordinary potential of low-dimensional systems in future technologies.

Sherbrooke, Canada Radouane En-nadir

Acknowledgements We would like to express our sincere gratitude to all the contributing authors and collaborators whose expertise and dedication have brought this project to life. Special thanks go to the research institutions and laboratories that provided a supportive environment for this interdisciplinary work. We also acknowledge the scientific community whose foundational work in quantum nanostructures continues to inspire new discoveries.

Contents

Fascinating New Generation Core–Shell Quantum Dots

Nabil Zeiri and Pınar Başer

1 Chapter 1

1.1 Introduction

When two or more semiconductor crystals with similar structures are brought together, the structure formed is called a heterostructure. It is assumed that the electronic structure and crystal potentials of the semiconductors forming the heterostructure do not change at the interface. The lattice constant and lattice structure of the heterostructures brought together should be close to each other. With this condition, it is assumed that the wave function of the electron does not change at the interface, i.e. the k wave vector does not change at the interface. The lattice constant is a measure of the suitability of the semiconductors when forming the heterostructure. The electron charge density of the semiconductors forming the structure should be the same on both sides of the interface. The Fermi energy depends on the charge density of the electron. At a temperature of $T = 0$ K, the Fermi energy in semiconductors is located right in the middle of the forbidden energy gap. The differences in electron density due to the difference between the lattice constants cause charge flow on the surface. The amount of this charge flow is negligible and this charge layer can be neglected. Heterostructures are divided into three types according to their magnification: Type I, Type II and Type III. Three different heterostructures are shown schematically in Fig. 1 [1].

As seen in Fig. 1, in Type I, the wide band gap of one semiconductor completely covers the smaller band gap of the other semiconductor. In other words, in this type

N. Zeiri
Laboratoire de La Matière Condensée Et Des Nanosciences (LMCN), Département de Physique, Faculté Des Sciences de Monastir, Université de Monastir, Monastir, Tunisie

P. Başer (✉)
Faculty of Science, Department of Physics, Sivas Cumhuriyet University, Sivas, Turkey
e-mail: pbaser34@gmail.com

R. En-nadir (ed.), *Quantum Systems at the Nanoscale*, Lecture Notes in Nanoscale Science and Technology 45, https://doi.org/10.1007/978-3-032-06645-9_1

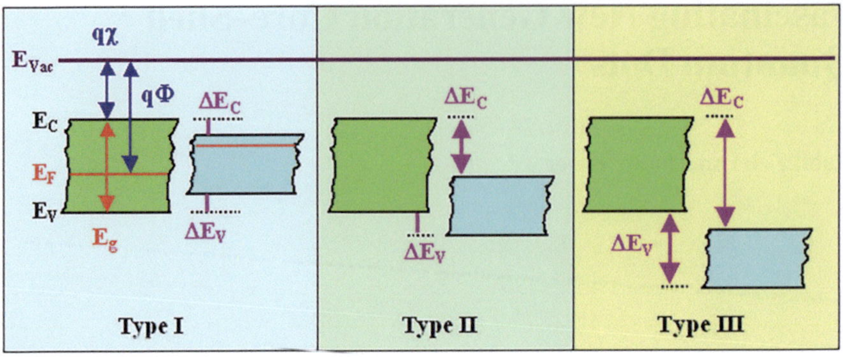

Fig. 1 Type I, Type II and Type III heterostructures

of heterostructure, $E_C(2) > E_C(1)$ and $E_V(2) < E_V(1)$. The discontinuities of the bands where there are two types of carriers such as electrons and holes are given by ΔE_c for electrons and ΔE_v for holes. When passing from the material with the smaller band gap to the material with the wider band gap, the charges at the edge gain energy. In the heterostructure for Type II, the band gaps of the two semiconductors overlap each other. One of the $\Delta E_{c/v}$ band discontinuities changes sign. This situation is caused by the movement of carriers from (1) to (2), in other words, the structure is no longer symmetrical. While some types of carriers gain energy (for example, electrons moving from right to left), others (holes) lose energy. In Type III, it is seen that the forbidden energy gaps of two different semiconductors no longer overlap. The reason for this is due to the movement of carriers, as explained in Type II.

Heterostructures or low-dimensional structures are named as quantum well (QWs), quantum well wires (QWWs) and quantum dots (QDs), according to the confinement dimension. While the electron movement in a QWs are restricted in one direction, the movement of an electron in a QWWs are restricted in both directions. Therefore, the degree of freedom (D_f) of the electron in the quantum wire is reduced by one degree compared to the electron in the quantum well. In the QD, since the electron's movement is restricted in 3-dimensions, the D_f is reduced to zero. If the number of degrees of freedom is denoted by D_f and the number of spatial confinement directions by D_c, then for all rigid systems, $D_f + D_c = 3$ can be written. The D_f and D_c values for the four cases shown in Table 1 are noteworthy [2].

Table 1 Representation of D_f and D_c for the bulk structure and three different low-dimensional structures

Structure	D_c	D_f
Bulk	0	3
Quantum well	1	2
Quantum wire	2	1
Quantum dot	3	0

This section will focus on the optoelectronic properties of core–shell quantum dots. Quantum dots, which are low-dimensional structures, are also defined as artificial atoms because they have discrete and interconnected electronic energy levels [3]. Quantum dots have dimensions on the scale of a few nanometers [4]. An artificial molecule could be obtained by combining two or more quantum dots [5]. The emission wavelengths of quantum dots, or their optoelectronic properties, can be altered by adjusting their size and shape [6]. Quantum dots have been used in many areas such as scanning tunneling microscopy [7], single-electron transistors [8], solar cells [9], lasers [10], second harmonic generation [11], third harmonic generation [12], quantum computing [13], biosensors[14], inkjet printing devices [15] and photodetectors [16] due to their different and superior properties compared to other low-dimensional structures.

Covering the quantum dot core with semiconductor material having different band gap creates a unique new quantum dot known as core–shell quantum dots [17–22]. CSQDs, which can be defined as a new model, exhibited quite different and unique properties. The shell surrounding the core increased the quantum efficiency by making the core more stable [23]. In addition, the electrical and optical properties of these CSQDs can be optimized by changing the shell thickness and the transition energies can be tuned by changing the size of these structures, which provides an additional improvement to the optical features [24, 25]. In addition, these new QDs have higher photoluminescence efficiency than old quantum dots because the carriers are surrounded by the shell material [17]. Optical and electronic properties of CSQDs could be tuned by doping [26]. It was observed that surrounding CSQDs with dielectric matrices (DM) containing different dielectric constants caused quite different changes on these structures [27–31]. These matrices surrounding CSQDs were found to increase structure stability, improve optical properties and provide many advantages [47]. Indeed, surrounding the CSQDs with Hafnium dioxide (HfO_2) [33], polyvinyl alcohol (PVA) [34], polyvinyl chloride (PVC) [35], silicon dioxide (SiO_2) [36], poly-methyl methacrylate (PMMA) [37], polyimide (PI) [38], is usually required to insure the QDs stabilization and enhances new optical properties. Some dielectric effects in reduced size materials can be listed as follows. Coulomb interactions between particles were found to be affected by the dielectric properties of the materials surrounding the low-dimensional structures. This effect is also called dielectric screening [39]. It was found that the PL measurements of single exciton, negative trion and biexciton particles in CdSe/CdS CSQDs varied with the dielectric environment [40].

Many studies have been done on the optoeelctronic feature of dielectric materials, electric fields (EF), and impurity atoms in core–shell quantum dots or surrounding conventional quantum dots [41–49]. The BE, photoionization cross section (PICS) and stark shift (SS) of a prolate ellipsoidal CSQDs containing donor impurities at the center and off-center under electric field were investigated. It was shown that the SS is strongly influced by the impurity location and the core and shell dimensions and can take positive or negative values. Furthermore, the core and shell dimensions, impurity location and EF also have a significant effect on the photoionization cross section [46]. In CdSe/ZnSe CSQDs structures, it was investigated how strain, structure size,

EF and MEH-PPV polymer matrix affect the energy values of the electron and the nonlinear optical reflection coefficient. It was observed that as the strain and electric field increased, the resonant amplitudes of the nonlinear optical reflection coefficient increased and the resonant frequencies shifted to the blue [49].

1.2 Hydrogenic Impurity

Since impurities originating from a single electron or a single hole form a structure similar to a hydrogen atom, these impurities in semiconductors are called hydrogenic impurities. The representation of the impurity atom in a bulk semiconductor is given in Fig. 2 [50].

The ionization energy (E_d) and bohr radius of a hydrogenic impurity in semiconductors can be calculated. For this, it is sufficient to take e^2 instead of to take e^2/ε and effective mass (m^*) instead of mass (m_0) in the ionization energy and bohr radius of the Hydrogen atom. Donor ionization energy and Bohr radius [51]:

$$\mathrm{E_d} = -\left(\frac{\mathrm{m}^*}{\mathrm{m_0}}\right)\frac{13.6}{\varepsilon^2}(\mathrm{eV}) \tag{1}$$

$$\mathrm{a_{B(d)}} = -0.53\varepsilon\left(\frac{\mathrm{m_0}}{\mathrm{m}^*}\right)\dot{\mathrm{A}} \tag{2}$$

The binding energy (BE) of a hydrogen dopant in a D-dimensional semiconductor is given by the following definition [4].

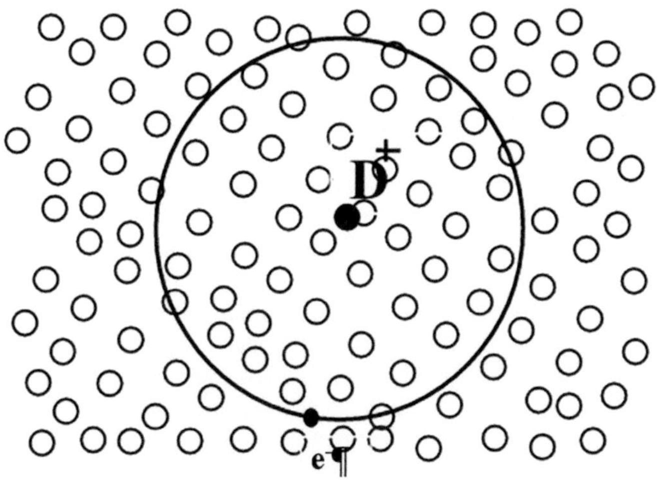

Fig. 2 Schematic representation of a donor impurity in a semiconductor bulk

$$E_b = \left(\frac{2}{D-1}\right)^2 \frac{R_B m_0}{\varepsilon_0^2} \tag{3}$$

Here R_B is the Rydberg constant and ε_0 is the static dielectric constant. ε is the dielectric constant of the semiconductor material. In the 2-dimensional (2D) case, the binding energy is four times larger than in the 3-dimensional case, while as can be seen from the definition of binding energy; in the one-dimensional case, the increase in binding energy is infinite. This is expected. Because as the dimension decreases, the motion of the electron is confined to a smaller dimension and spends most of its time around the dopant. Therefore, the binding energy of the electron is expected to be larger in lower dimensions.

1.3 *Dielectric Effect*

In CSQD structures, one of the parameters used to change the electronic and optical parameters is to embed these structures in dielectric matrix elements. Three different cases can be encountered for this dielectric matrix element: First, the dielectric function of CSQDs, denoted by ε_{in}, is larger than the dielectric function of the matrix ε_{out} in which the structure is embedded ($\varepsilon_{in} > \varepsilon_{out}$), second, the matrix ε_{in} _ is smaller than ε_{out} ($\varepsilon_{in} < \varepsilon_{out}$), or finally $\varepsilon_{in} = \varepsilon_{out}$. This dielectric mismatch affects the confinement potential due to self-energy polarization.

The effects of these three situations can be summarized as follows:

1. If $\varepsilon_{in} < \varepsilon_{out}$, the resonance peaks of PICS shift to lower energies and the PICS resonant amplitudes increase.
2. $\varepsilon_{in} < \varepsilon_{out}$, PICS resonance peaks shift to higher energies and PICS resonant amplitudes decrease.
3. If $\varepsilon_{in} = \varepsilon_{out}$, out, the limiting band offset alignment is reduced to V_{0e} In this case, the system is embedded in a dielectrically homogeneous medium that does not affect the optical properties.

1.4 *Semiconductor Compounds*

1.4.1 CdS

Cadmium sulfide (CdS) is a II–VI compound semiconductor with a direct band gap Eg = 2.42 eV [52]. It has very good thermal and chemical stability. It is used in the design of many optoelectronic devices such as solar cells, light emitting diodes and lasers. CdS has three types of crystal structures: cubic zinc blende (found in the mineral Hawleyite), hexagonal wurtzite (found in the mineral Greenockite), and cubic closed packed rock salt phase. In Fig. 3a, b and c, the unit cell of the wurtzite

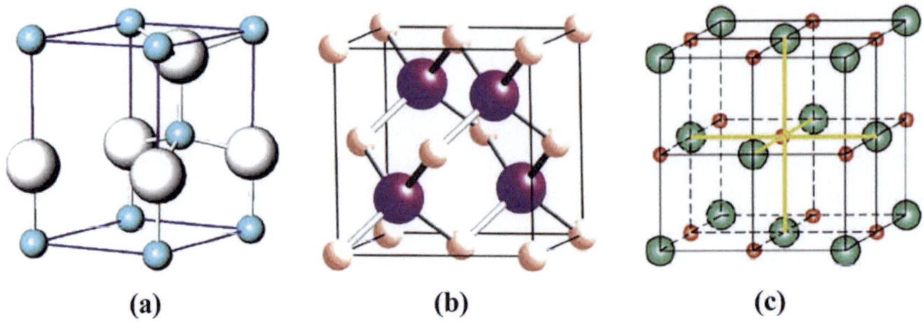

Fig. 3 Unit cell for crystal structure of CdS: **a** wurtzite (hcp), **b** zinc mixture and **c** cubic close-packed (ccp)

(hcp), zinc mixture and cubic close-packed (ccp) crystal structure of CdS is presented, respectively.

The Greenockite structure is a hexagonal, yellowish crystal. It is a rare mineral and the only true ore of cadmium. It is a highly sought-after mineral due to its beautiful color and crystal structure. The structure of Greenockite consists of SCd_4 tetrahedrons stacked in a layered structure where every other layer is exactly the same in the AB AB AB hexagonal arrangement. Greenockite crystals are usually six-sided pyramids. The other polymorph of Cadmium Sulfide, the cubic hawleyite structure, is a rare sulfide mineral in the sphalerite group [53].

Colloidal dispersions of CdS semiconductor nanoparticles can exhibit fluorescence color change depending on the size of the particle. CdS nanoparticles exhibit quantum size effect because the size of the CdS particles is directly related to the absorption wavelength.

1.4.2 ZnSe

Zinc selenide (ZnSe) is a semiconductor compound of group II–VI. It has direct energy gap and its value is $Eg = 2.7$ eV [54]. In addition to the wide band gap, ZnSe has very good electronic transport properties, high linear and nonlinear refractive indices, and good optical transparency over a wide range in the visible spectrum. It also has a large exciton binding energy of 21 meV [54]. ZnSe is used in lenses, lasers, light emitting diodes, window materials for solar cells, and infrared cameras [55, 56]. The crystal structure of ZnSe is hexagonal wurtzite or cubic blende. In Fig. 4a, b the unit cell of the wurtzite (hcp), zinc blende crystal structure of ZnSe is presented, respectively [57].

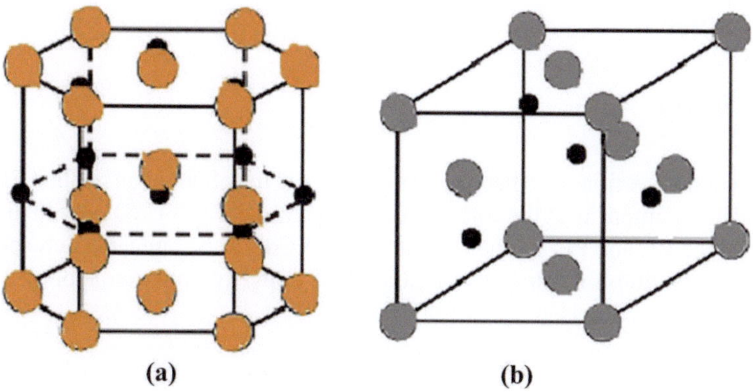

Fig. 4 Unit cell for crystal structure of ZnSe: **a** wurtzite, **b** zinc blende

1.4.3 ZnS

Zinc sulfide (ZnS) is a semiconductor compound of group II–VI. At room temperature, the direct energy gap Eg = 3.54 eV in the cubic structure of ZnS, while the direct band gap Eg = 3.91 eV in the hexagonal structure [58]. ZnS can be doped as an n-type semiconductor or a p-type semiconductor. ZnS can be transparent over a wide energy range, having a very high transmittance from visible wavelengths to just over 12 μm. Because of these properties, it is used in many optoelectronic devices, including solar cells, liquid crystal (oil panel) displays, light emitting diodes, and sensors for visible and infrared optics, nanowires nanoribbons, nanotubes and biomedical fields [59–63]. ZnSe can have two forms of crystal structure, namely wurtzite and zinc blende. In Fig. 5a and b the unit cell of the wurtzite, zincblende crystal structure of ZnS is presented, respectively [64].

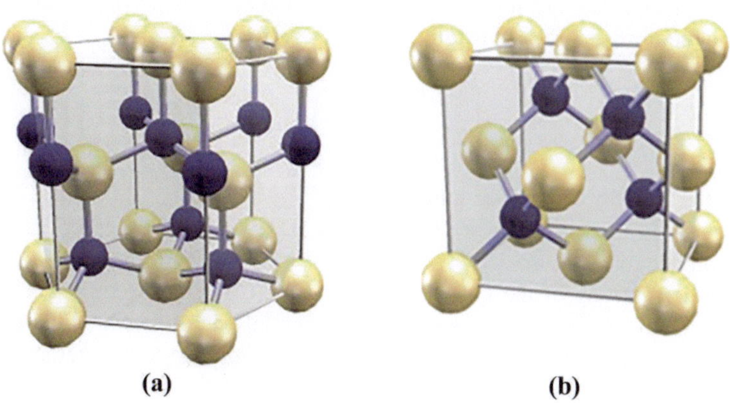

Fig. 5 Unit cell for crystal structure of ZnS: **a** wurtzite, **b** zinc blende

2 Chapter 2

2.1 Theory

The aim of this section is to theoretically investigate the effect of an applied EF, QDs size and donor impurity at the core of CdS/ZnSe and CdS/ZnS CSQDs with different D.M.s on the BE and PICS. The Hamiltonian for CdS/ZnSe and CdS/ZnS spherical CSQDs containing a single electron and donor impurity under an applied electric field in the z direction is given by [65]:

$$\hat{H}\Psi_{n,l}(\vec{r}) = E_{n,l}(\vec{r})\Psi_{n,l}(\vec{r}) \tag{4}$$

$$\hat{H} = \frac{\hbar-^{2}}{2}\vec{\nabla}\left(\frac{1}{m^{*}_{c,s}}\vec{\nabla}\right) + V_c(\vec{r}) + \sum(R_s) + U(\vec{r}) + e.F.r.\cos(\theta) \tag{5}$$

$V_c(\vec{r})$ is the spatial confinement potential for CdS/ZnSe and CdS/ZnS CSQDs structures and is shown for each region as [66, 67]:

$$\mathrm{V_c}(\vec{\mathrm{r}}) = \begin{cases} 0 & 0 \leq \mathrm{r} \leq \mathrm{R_c} \\ \mathrm{V_{0e}} & \mathrm{R_c} \leq \mathrm{r} \leq \mathrm{R_s} \\ \infty & \mathrm{r} \geq \mathrm{R_s} \end{cases} \tag{6}$$

where m^{*}_{c} and m^{*}_{s} define the effective mass in the core and shell regions, respectively and is given with its definition for each region.

$$m^{*}_{c,s} = \begin{cases} m^{*}_{c} & r < R_c \\ m^{*}_{s} & R_c < r \leq R_s \end{cases} \tag{7}$$

And $\sum(\mathrm{R_s})$ designate the self-polarization potential induced by the external dielectric medium, expressed as [68]:

$$\sum(R_s) = \frac{e^2}{4\pi\varepsilon_0 R_s}\left[\frac{1}{2}\left(\frac{1}{\varepsilon_{\text{out}}} - \frac{1}{\varepsilon_{\text{in}}}\right) + \frac{0.466}{\varepsilon_{\text{in}}}\left(\frac{\varepsilon_{\text{in}} - \varepsilon_{\text{out}}}{\varepsilon_{\text{in}} + \varepsilon_{\text{out}}}\right)\right] \tag{8}$$

ε_0 is the permittivity in vacuum, $\varepsilon_{in} = \sqrt{\varepsilon_{CdS}.\varepsilon_{ZnSe}}$ or $\varepsilon_{in} = \sqrt{\varepsilon_{CdS}.\varepsilon_{ZnS}}$ are the permittivity of CdS/ZnSe or CdS/ZnS CSQDs. The values appropriate to the semiconductor material the CSQDs structure is composed of are used. ε_{out} is the dielectric permittivity of the D.M. surrounding the CSQDs structure. The energy $U(\vec{r})$ is defined as the sum of the Coulomb potential energy between the electron and the impurity and the potential energy of the electron due to its interaction with the polarization charges [69].

$$U(\vec{r}) = U_{impurity} + U_{polarizationcharges} \tag{9}$$

$$U_{impurity} = -\frac{e^2}{4\pi \varepsilon_{in} \left| \vec{r}_e - \vec{r}_{imp} \right|} \tag{10}$$

$$U_{polarizationcharges} = -\frac{e^2(\varepsilon_{in} - \varepsilon_{out})}{4\pi \varepsilon_{in}\varepsilon_{out} R_s} \tag{11}$$

In the CSQDs structure, when no external electric field is applied, the dielectric is not embedded with matrix elements, and the structure does not contain impurity atoms, the wave functions of the Hamiltonian of the system consisting of a single electron are defined as:

$$\Psi^0_{n,l}(r,\theta,\varphi) = R^0_{n,l}(r).Y_{l,m}(\theta,\varphi) \tag{12}$$

In this wave function, $R^0_{n,l}(r)$ represents the radial part and $Y_{l,m}(\theta,\varphi))$ represents the spherical harmonics [70].

$$R^0_{n,l}(r) = \begin{cases} A_1 J_l\left(k_{nl,1}.r\right) 0 \le r \le R_c \\ B_1 h_l^+\left(ik_{nl,1}.r\right) + B_2 h_l^-\left(ik_{nl,2(ZnSe)} or (ik_{nl,2(ZnS)}).r\right) R_c \le r \le R_s \\ 0 r \ge R_s \end{cases} \tag{13}$$

Normalization constants A_1, B_1 and B_2 can be determined from these boundary conditions [70, 71]. The coefficients $k_{nl,1}$ and $k_{nl,2}$ depending on the electronic energy value in Eq. (13) are given by Eqs. (14) and (15).

$$k_{nl,1} = \frac{\sqrt{2m^*_{CdS}(E_{n,l} - \sum(R_s))}}{\hbar^2};\ k_{nl,2(ZnSe)} = \frac{\sqrt{2m^*_{ZnSe}(V_c + \sum(R_s) - E_{n,l})}}{\hbar^2} \tag{14}$$

$$\mathrm{k_{nl,1}} = \frac{\sqrt{2\mathrm{m}^*_{\mathrm{CdS}}(\mathrm{E_{n,l}} - \sum(\mathrm{R_s}))}}{\hbar^2};\ \mathrm{k_{nl,2(ZnS)}} = \frac{\sqrt{2\mathrm{m}^*_{\mathrm{ZnS}}(\mathrm{V_c} + \sum(\mathrm{R_s}) - \mathrm{E_{n,l}})}}{\hbar^2} \tag{15}$$

In order to find the wave function of the Hamiltonian in Eq. (5) containing perturbation terms (impurity, the electric field, dielectric matrix), a new trial wave function was obtained by multiplying the wave function defined by $R^0_{n,l}(r)$, with a function containing the λ variational parameter to account for the impurity effect and the η variational parameter to account for the electric field effect, as required by the variational method. The trial function $R_{\lambda,\eta}(r)$ associate to the impurity ground state is taken from [71]:

$$\mathrm{R}_{\lambda,\eta}(\mathrm{r}) = \mathrm{R}^0_{\mathrm{n,l}}(\mathrm{r})\mathrm{e}^{-\lambda r}\mathrm{e}^{-\eta r\cos\theta} \tag{16}$$

According to perturbation theory, the ground state electronic energy level E of the system with hydrogenic impurity can be obtained by minimizing the total energy with respect to λ and η [71, 73].

$$E = \min_{\lambda,\eta} \frac{\Psi_{n,l}(r,\theta,\varphi)|\hat{H}|\Psi_{n,l}(r,\theta,\varphi)}{\Psi_{n,l}(r,\theta,\varphi)|\Psi_{n,l}(r,\theta,\varphi)} \tag{17}$$

The BE is defined as the difference between the ground state energies without and with impurity [74]:

$$E_b = E_0 - E \tag{18}$$

The general formula of PICS was defined as [75, 78]:

$$\sigma(\hbar\omega) = \left[\left(\frac{F_{eff}}{F_0}\right)^2 \frac{n_r}{\varepsilon_{in}}\right] \frac{4\pi^2}{3} \beta_{FS} \hbar\omega \sum_f |M_{if}|^2 \delta(\Delta E_{if} - \hbar\omega) \tag{19}$$

In the above equation, n_r is the refractive index of CSQDs, $\hbar\omega$. represents the incident photon energy. Where the ratio F_{eff}/F_0 is defined as the ratio of the incident effective field in the medium to the average field. This ratio is equal to 1 since it does not affect the PICS shape. $\beta_{FS} = \frac{e^2}{4\pi\varepsilon_0\hbar c}$ is the fine structure parameters, ΔE_{if} represent the difference in energy.

$M_{if} = F\Psi_i|r|\Psi_f$ designate the dipole matrix element, Ψ_i and Ψ_f are the initial and final eigenwave functions with and without impurity givens respectively. F is a dimensionless factor describing the discontinuity of EF inside and outside the CSQDs given by [79]:

$$F = \frac{3\varepsilon_{out}}{\varepsilon_{in} + 2\varepsilon_{out}} \tag{20}$$

And $\delta\big(\Delta E_{if} - \ hbar\omega\big)$ is the Lorentzian function given by:

$$\delta(\Delta E_{if} - \hbar\omega) = \frac{\hbar\Gamma}{\pi\left[(\hbar\omega - (\Delta E_{if}))^2 + (\hbar\Gamma)^2\right]} \tag{21}$$

where $\Gamma = \frac{1}{\tau}$ is the hydrogenic impurity line width.

3 Chapter 3

3.1 Results and Discussion

In this section, we calculate and interpret the impurity binding energy and PICS of a spherical CdS/ZnSe CSQDs dispersed in SiO_2, PVC and PVA D.M. and a spherical CdS/ZnS CSQDs dispersed in PI, PVC and PVA D.M. in Fig. 1a and b, respectively. The dielectric constant of CdS/ZnSe CSQDs is $\varepsilon_{in}(\text{CdS/ZnSe}) = \sqrt{\varepsilon_{CdS}.\varepsilon_{ZnSe}}$

= 5.36 [80]. The dielectric constant of CdS/ZnS CSQDs is $\varepsilon_{in}(\text{CdS/ZnS}) = \sqrt{\varepsilon_{CdS} \cdot \varepsilon_{ZnS}} = 8.59$ [81]. Fig 1a and b [82, 83] show the variation of the CdS/ZnSe and CdS/ZnS spherical CSQDs structure and the confinement potential with the radial distance, respectively. V_{0e} is the electron confinement potential due to the conduction band discontinuity. The physical parameters used in our calculations are listed in Tables 2 and 3 (Fig. 6).

In Fig. 7a and b [82, 83], the variation of BE with respect to Rc/Rs is presented for PVA, PVC and SiO_2 D.M.s (CdS/ZnSe) and PI, PVC and PVA D.M.s (CdS/ZnS), respectively.

In Fig. 7a and b, to the increase in the confinement effects. It was observed that E_b suddenly started to decrease when this ratio decreased after this critical value of R_c/

Table 2 Some parameters for CdS, ZnS and ZnSe semiconductors

Materials	m^*/m_0	ε	V_{0e} (eV) CdS/ZnSe	V_{0e} (eV) CdS/ZnS
CdS	0.18 [79]	8.3 [79]		
ZnS	0.42 [47]	8.9 [79]		0.897 [47]
ZnSe	0.15 [80]	5.23 [80]	0.80 [47]	

Table 3 D.M.'s and dielectric constants used in calculations

D.M	Dielectric constant (ε_{out})
PI	2.6 [83]
PVC	4 [83]
PVA	14 [47]
SiO_2	3.9 [83]

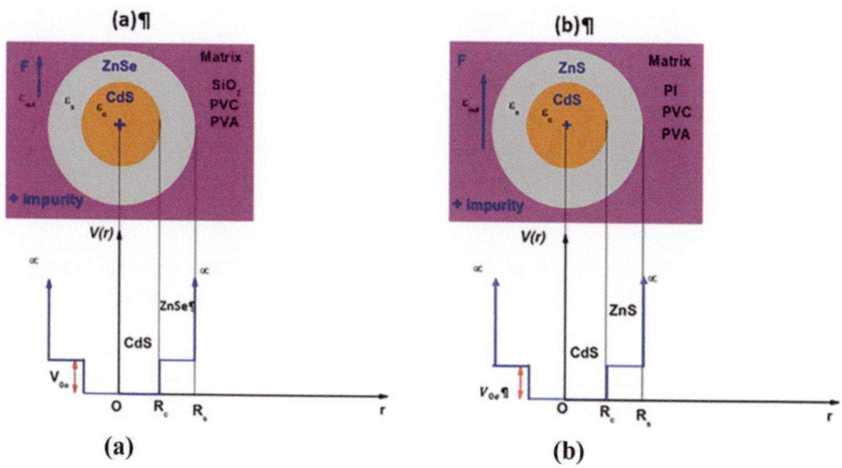

Fig. 6 **a** Schematic plot of CdS/ZnSe CSQDs and **b** CdS/ZnS CSQDs coated by different matrices

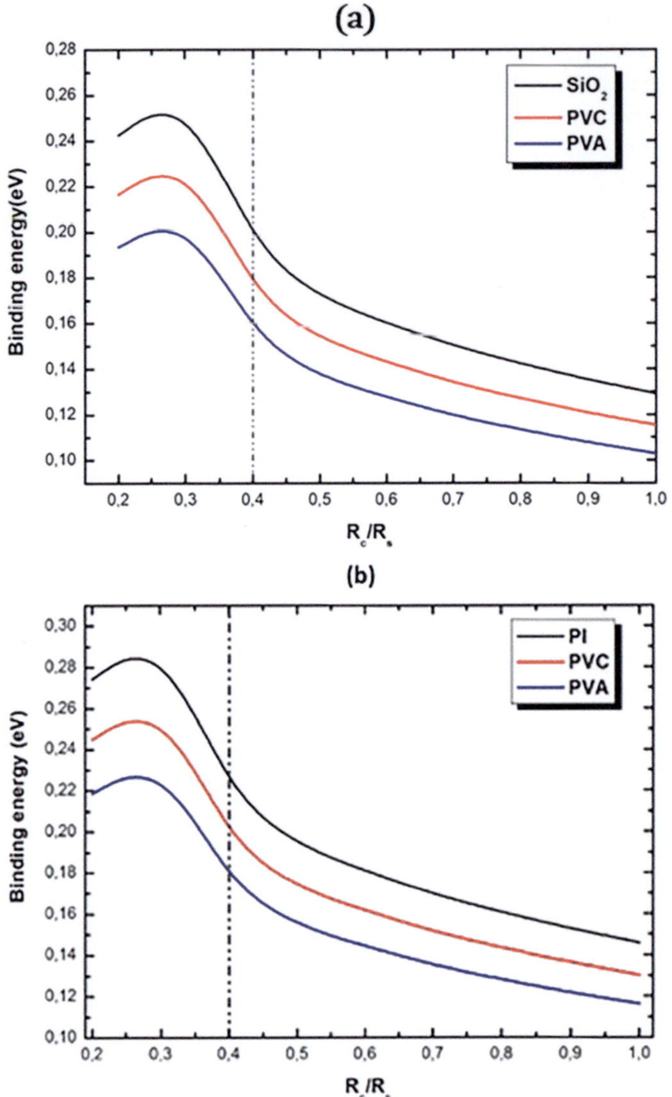

Fig. 7 **a** For PVA, PVC and SiO_2, D.M.s (CdS/ZnSe CSQDs), BE variation with respect to R_c/R_s **b** For PI, PVC and PVA D.M.s (CdS/ZnS CSQDs), BE variation with respect to R_c/R_s

R_S. Because in this case, the electron leaks from the core to the shell region and starts to move away from the impurity atom. In the dielectric material, as the dielectric constant increases, the Coulomb interaction decreases even more and therefore the dEb values also start to decrease. For example, among these four different D.M., the highest dielectric constant is in the PVA matrix. For CdS/ZnSe CSQDs and CdS/ZnS CSQDs, the binding energy in the PVA matrix takes its smallest value.

Figure 8a and b show the variation of BE with the EF for CdS/ZnSe and CdS/ZnS CSQDs structures at Rc/Rs = 0.4 for different D.M.s, respectively [82, 84]. It is seen that BE decreases as EF increases. Because, as the electric field increases, the Coulomb interaction between electrons and impurities decreases.

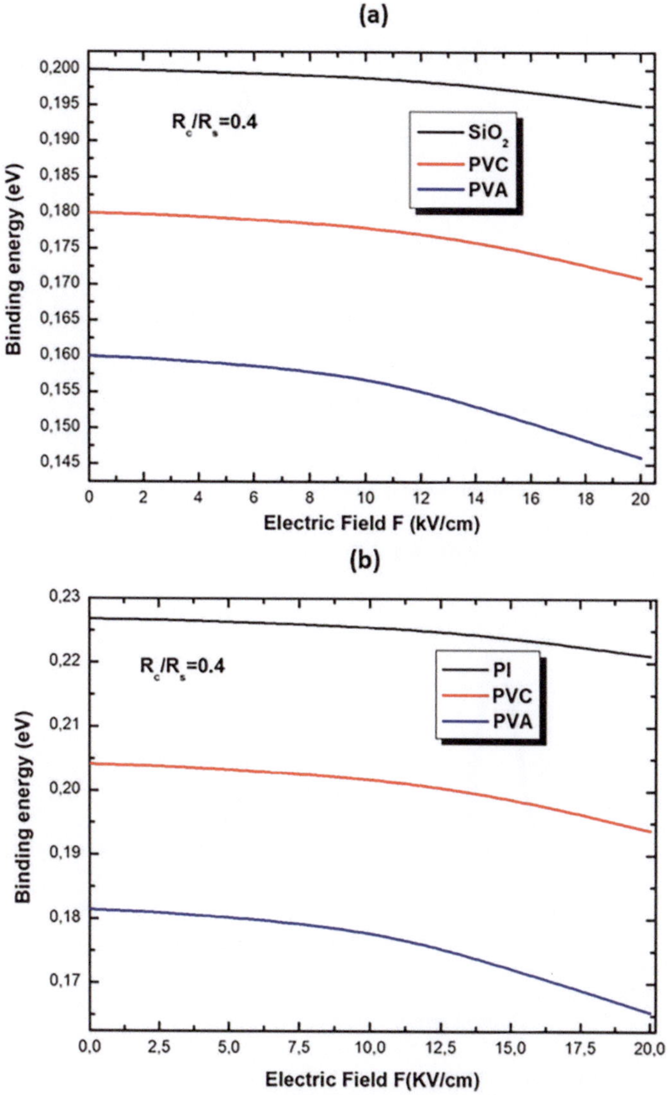

Fig. 8 **a** For PVA, PVC and SiO_2, D.M.s (CdS/ZnSe CSQDs), BE variation with respect to EF **b** For PI, PVC and PVA D.M.s (CdS/ZnS CSQDs), BE variation with respect to EF

In the absence of electric field, the variation of PICS versus incident photon energy for CdS/ZnSe CSQDs, PVA, PVC and SiO_2 embedded D.Ms is shown in Fig. 9a for Rc/Rs = 0.4 and the variation of PICS for CdS/ZnS CSQDs, PVA, PVC and PI, D.M.s is shown [82] in Fig. 9b. Here, two cases are observed depending on ε_{in} of CSQDs.

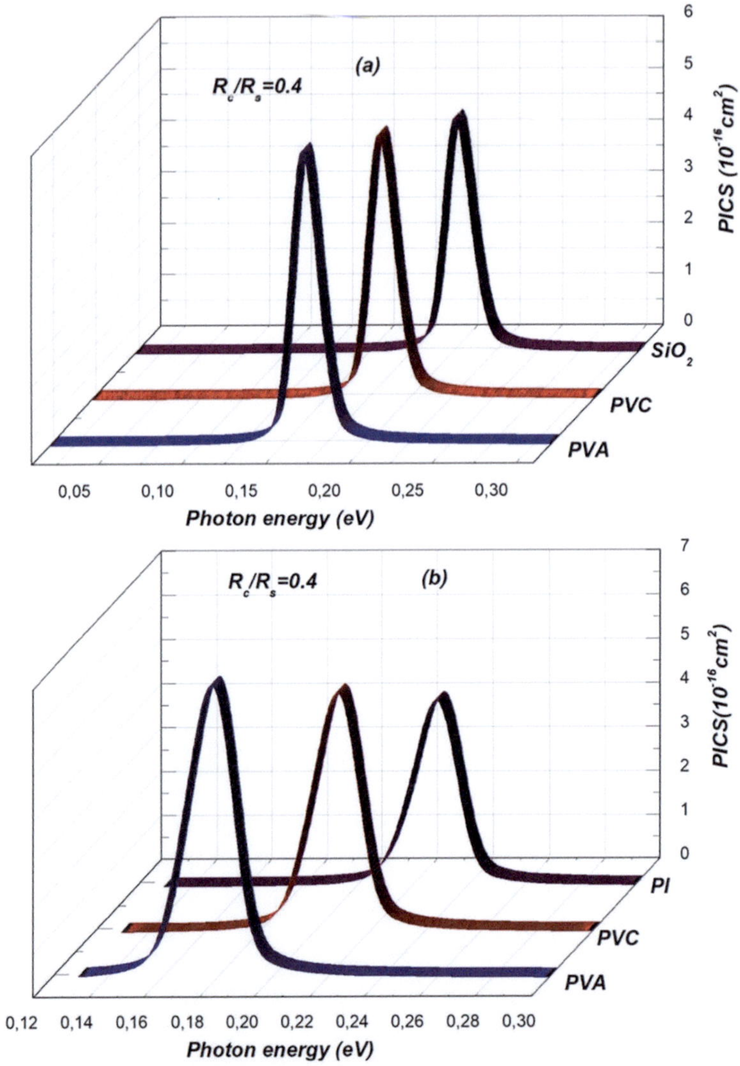

Fig. 9 **a** For PVA, PVC and SiO2, D.M.s (CdS/ZnSe CSQDs), PICS variation depending on incident photon energy **b** for PI, PVC and PVA D.M.s (CdS/ZnS CSQDs), PICS variation depending on incident photon energy

Here, two cases were observed depending on the ε_{in} of CSQD. For CdS/ZnSe CSQDs, the ε_{in} in value is larger than ε_{out} (PVA) and ε_{out} (SiO_2). In other words, if $\varepsilon_{in} > \varepsilon_{out}$, the local field factor F < 1 will be, in which case the PICS resonant frequencies will be blue-shifted and will cause a weaker PICS. If $\varepsilon_{in} < \varepsilon_{out}$, in other words, when the DM matrix is PVA, the local field factor F > 1 will be, which will cause an increase in the transition matrix element, which will give a stronger PICS. Also, in this case, the PICS resonant frequencies will be red-shifted. For CdS/ZnS CSQDs, the value of ε_{in} is greater than the value of ε_{out} (PI). In other words, if $\varepsilon_{in} > \varepsilon_{out}$, the local field factor F < 1 will be blue-shifted, resulting in a weaker PICS. If $\varepsilon_{in} < \varepsilon_{out}$, that is, when the DM matrix is PVA and PVC, the local field factor F > 1 will be stronger, resulting in a stronger PICS. Also, in this case, the PICS resonant frequencies will be red-shifted.

In Fig. 10a and b [82], PICS changes with respect to incident photon energy at different electric field values for CdS/ZnSe and CdS/ZnS CSQDs structures, respectively, for PVA matrix and Rc/Rs = 0.4 are presented.

3.2 Conclusion

In this section, we investigated the electronic and some optical properties for CSQDs obtained from two different materials, CdS/ZnSe and CdS/ZnS. Using the effective mass approach and variational technique, we calculated and interpreted the BE and PICS values of a hydrogenic impurity at the center of CSQDs under the applied external electric field and when these structures were embedded in different dielectric matrix elements. Since the confinement potential heights of CSQDs particles were different from each other, it was found that BE values were different for the same matrix elements (PVA, PVC) due to different electronic energy values. The obtained results showed that the dielectric material can be used as a tuning parameter, like the electric field, in tuning the optical and electronic properties. It was also seen that the EF and the external dielectric matrix surrounding the structures were effective in improving the PICS values. It was concluded that the electric enclosure and electric field were quite suitable for obtaining device optima in new device designs.

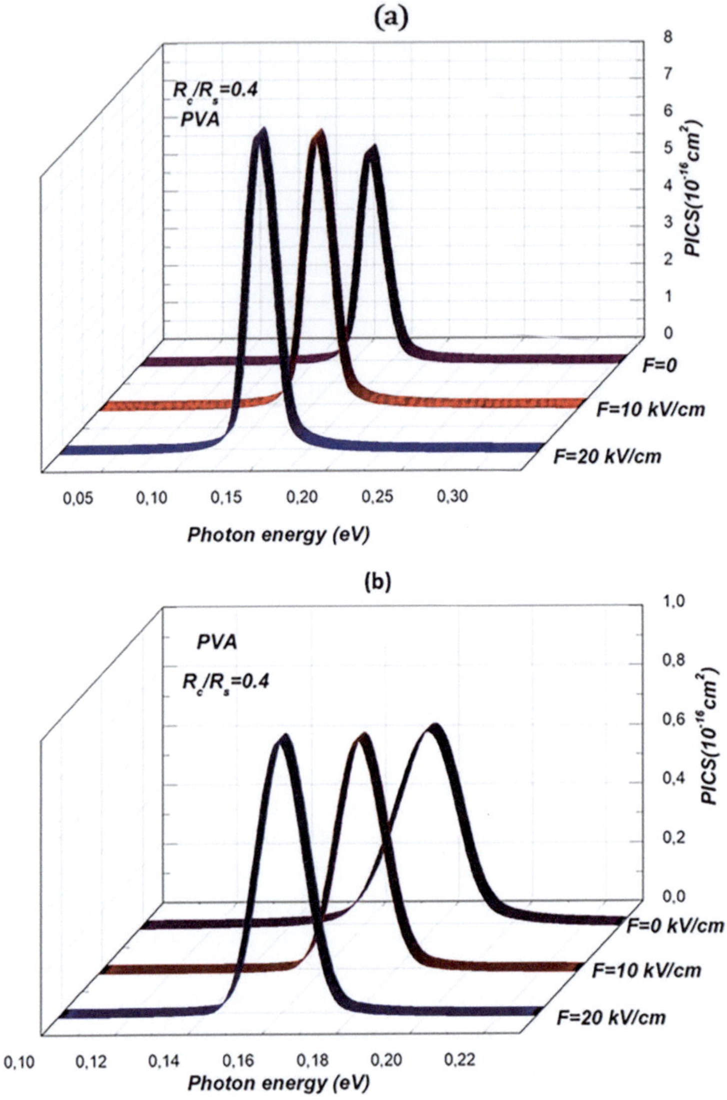

Fig. 10 For the PVA dielectric matrix, the change of PICS with the incident photon energy at F = 0, 10 and 20 (CdS/ZnSe CSQDs) (**a, b**) For the PVA dielectric matrix, the change of PICS with the incident photon energy at F = 0, 10 and 20 (CdS/ZnS CSQDs)

References

1. M.Y. Wang, Dislocation et relaxation des contraintes aux interfaces entre semiconducteurs III–V à large différence de paramètres de maille, Mater. Sci. [cond-mat. mtrl-sci]. Université de Caen (2012)

2. P. Baser, Electronic properties of low dimensional systems: the hydrogenic impurities and excitonic binding energies in cylindrical $Ga_{1-x}Al_xAs/GaAs$ quantum wires under an externally applied magnetic field. Doctorate Thesis (2007)
3. B. Boyacioglu, A. Chatterjee, Dia-and paramagnetism and total susceptibility of GaAs quantum dots with Gaussian confinement. Phys. E **44**, 1826–1831 (2012)
4. O.E. Semonin, J.M. Luther, M.C. Beard, Quantum dots for next-generation photovoltaics. Mater. Today **15**, 508–515 (2012)
5. J. Cui, Y.E. Panfil, S. Koley, D. Shamalia, N. Waiskopf, S. Remennik, I. Popov, M. Oded, U. Banin, Colloidal quantum dot molecules manifesting quantum coupling at room temperature. Nat. Commun. **10**, 5401 (2019)
6. C.B. Murray, C.R. Kagan, M.G. Bawendi, Synthesis and characterization of monodisperse nanocrystals and close-packed nanocrystal assemblies. Annu. Rev. Mater. Res. **30**, 545–610 (2000)
7. E.P.A.M. Bakkers, Z. Hens, A. Zunger, A. Franceschetti, L.P. Kouwenhoven, L. Gurevich, D. Vanmaekelbergh, Shell-tunneling spectroscopy of the single-particle energy levels of insulating quantum dots. Nano Lett. **1**, 551–556 (2001)
8. S.K. Sahu, K. Mazumdar, State-of-the-art Analysis of Single Electron Transistor, in *IEEE 3rd International Conference on Electrical Power and Energy Systems (ICEPES)*. (MANIT Bhopal (MP), India, 2024) pp.1–6
9. M. Rahn Kim, D. Ma, Quantum-dot-based solar cells: recent advances, strategies, and challenges. J. Phys. Chem. Lett. **6**, 85–99 (2015)
10. W. Wang, S. Ding, Z. Wang, F. He, S. Zhao, T. Wang, J. Zhang, X. Xu, Y. Yao, H. Huang, F. Grillot, J. Duan, Intensity noise reduction in quantum dot comb laser by lower external carrier fluctuations. Opt. Lett. **49**, 5007–5010 (2024)
11. X. Yan, X. Li, Y. Cai, X. Wang, Nonlinear optical second harmonic generation characteristics in cylindrical GaAs/Ga1−ηAlη as quantum dots. Phys. Status Solidi B **261**(8), 2400136 (2024)
12. K. Hammouda, N. Yahyaoui, N. Zeiri, A. Ed-Dahmouny, P. Baser, A. Sali, M. Said, M.N. Murshed, Third-order nonlinear optical susceptibility of hydrogenic impurity in Ge/$Si_{0.15}Ge_{0.85}$ spherical core/shell quantum dots under electric field. Solid State Commun. **383**, 115480 (2024)
13. D. Loss, D.P. DiVincenzo, Quantum computation with quantum dots. Phys. Rev. A **57**, 120 (1998)
14. F. Ma, C.-C. Li, C.-Y. Zhang, Development of quantum dot-based biosensors: principles and applications. J. Mater. Chem. B **39**, 6173–6190 (2018)
15. C. Wei, B. Xu, M. Zhang, Z. Su, J. Gu, W. Guo, X. Gao, W. Su, Z. Cui, S. Jeon, Z. Fan, H. Zeng, Highly ordered inkjet-printed quantum-dot thin films enable efficient and stable QLEDs with EQE exceeding 23%. eScience **4**, 100227 (2024)
16. D. Wu, G. Xu, J. Tan, X. Wang, Y. Zhang, L. Ma, W. Chen, K. Wang, Nanophotonic structures energized short-wave infrared quantum dot photodetectors and their advancements in imaging and large-scale fabrication techniques. Nanoscale **17**, 8239–8269 (2025)
17. D. Vasudevan, R.R. Gaddam, A. Trinchi, I. Cole, Core–shell quantum dots: properties and applications. J. Alloy. Compd. **636**, 395–404 (2015)
18. A. Kumar, D. Kumar, Core-shell quantum dots: a review on classification, materials, application, and theoretical modeling. J. Alloy. Compd. **924**, 166508 (2022)
19. O.I. Micic, B.B. Smith, A.J. Nozik, Core-shell quantum dots of lattice-matched $ZnCdSe_2$ shells on InP cores: experimentand theory. J. Phys. Chem. B **104**, 12149–12156 (2000)
20. J. Li, L.-W. Wang, First principle study of core/shell structure quantum dots available to purchase. Appl. Phys. Lett. **84**, 3648–3650 (2004)
21. N. Jin, Y. Nagaoka, Z. Liu, R. Li, O. Chen, Pressure-modulated energy transfer dynamics in Mn^{2+}-Doped CdS/ZnS core/shell quantum dots. J. Am. Chem. Soc. **147**, 7965–7973 (2025)
22. S. Mathew, A.D. Saran, B. Singh Bhardwaj, S. Ani Joseph, P. Radhakrishnan, V.P. Nampoori, C.P. Vallabhan, J.R. Bellare, Size dependent optical properties of the CdSe-CdS core-shell quantum dots in the strong confinement regime. J. Appl. Phys. **111**, 074312 (2012)

23. R. Shwetharani, N. Vignesh, M.S. Jyothi, R. Geetha Balakrishna, Review on recent advances of core-shell structured lead halide perovskites quantum dots. J. Alloys Compounds **834**, 155246 (2020)
24. X. Feng, G. Xiong, X. Zhang, H. Gao, Third-order nonlinear optical susceptibilities associated with intersubband transitions in CdSe/ZnS core–shell quantum dots. Phys. B Condensed Matter **383**, 207–212 (2006)
25. A. Sahu, D. Kumar, Core-shell quantum dots: a review on classification, materials, application, and theoretical modeling. J. Alloy. Compd. **924**, 166508 (2022)
26. P. Malik, R. Thareja, J. Singh, R. Kakkar, II-VI core/shell quantum dots and doping with transition metal ions as a means of tuning the magnetoelectronic properties of CdS/ZnS core/shell QDs: A DFT study. J. Mol. Graph. Model. **111**, 108099 (2022)
27. W. Li, K. Li, X. Zhao, C. Liu, Coudert FX defective nature of CdSe quantum dots embedded in inorganic matrices. J. Am. Chem. Soc. **144**, 11296–11305 (2022)
28. Y. Zhao, Co-precipitated Ni/Mn shell coated nano Cu-rich core structure: a phase-field study. J. Mater. Res. Technol. **21**, 546–560 (2022)
29. Y. Qiu, M. Shi, X. Guo, J. Li, J. Wu, Y. Zhou, H. Sun, Y. Hang, X. Li, Y. Li, Sensitivity improvement in the measurement of minor components by spatial confinement in fiber-optic laser-induced breakdown spectroscopy. Spectrochim. Acta Part B: Atomic Spectroscopy. **209**, 106800 (2023)
30. K. Hasanirokh, A. Naifar, Mathematical modeling of core-multishell quantum dot with capped oxides: a comprehensive study. Plasmonics **19**, 2317–2332 (2024)
31. Z. Zeng, C.S. Garoufalis, A.F. Terzis, S. Baskoutas, Linear and nonlinear optical properties of ZnO/ZnS and ZnS/ZnO core shell quantum dots : Effects of shell thickness, impurity, and dielectric environment. J. Appl. Phys. **114**, 023510 (2013)
32. A. Cherni, N. Zeiri, N. Yahyaoui, M. Said, Consequences of dielectric mismatch on linear and third order nonlinear optical properties for CdS/ZnSe core/shell QD-matrix. Chem. Phys. **539**, 110947 (2020)
33. A.I. Ankhi, M.R. Islam, M.T. Hasan, E. Hossain, Projected performance of InGaAs/GaAs quantum dot solar cells: effects of cap and passivation layers. IEEE Access **8**, 212339–212350 (2020)
34. Z.K. Heiba, M.B. Mohamed, N.G. Imam, N.Y. Mostafa, Optical and electrical properties of quantum composite of polyvinyl alcohol matrix with CdSe quantum dots. Colloid Polym. Sci. **294**, 357–365 (2016)
35. R.R. Shamilov, V.I. Nuzhdin, V.F. Valeev, Yu.G. Galyametdinov, A.L. Stepanov, Photoluminescence of composite fılms of poly(n-vınylcarbazole) wıth CdCe/CdCcore/shell quantum dots located near the layer of sılver nanopartıcles on a dielectric material. J. Appl. Spectrosc. **82**, 773–778 (2015)
36. A. Arfaoui, T.M. Althagafi, Quadratic Electro-Optic Effect and Electro-Absorption Process in Cdse/Cds Core/Shell Quantum Dots. papers.ssrn.com, pp.1–25 (2025)
37. A. Ed-Dahmouny, N. Zeiri, R. Arraoui, N. Es-Sbai, M. Jaouane, A. Fakkahi, A. Sali, K. El-Bakkari, C.A. Duque, The third-order nonlinear optical susceptibility in an ellipsoidal core-shell quantum dot embedded in various dielectric surrounding matrices. Phys. E Low-dimensional Syst. Nanostruct. **153**, 115784 (2023)
38. W. Deng, G. Ren, W. Wang, W. Cui, W. Luo, Enhanced dielectric properties and thermostability in polyimide composites with core-shell structured polyimide@BaTiO3 nanoparticles. High Perform. Polym. **33**, 1–9 (2021)
39. F. Chen, J. Xia, D.K. Ferry, N. Tao, Dielectric screening enhanced performance in graphene FET. Nano Lett. **9**, 2571–2574 (2009)
40. M. Zhu, J. Zhou, Z. Hu, H. Qin, X. Peng, Effects of local dielectric environment on single-molecule spectroscopy of a CdSe/CdS Core/Shell quantum dot. ACS Photon. **5**, 4139–4146 (2018)
41. E. Sadeghi, Impurity binding energy of excited states in spherical quantum dot. Phys. E **41**, 1319–1322 (2009)

42. I. Maouhoubi, R. En-nadir, S. Maouhoubi, I. Zorkani, A.O.T. Hassani, A. Jorio, The effects of lateral electric field, effective-mass position-dependent, and dimension on the donor-impurity binding energy and electron-impurity distance in GaAs-based quantum disk. Phys. E **148**, 115630 (2023)
43. C.C. Yang, L.C. Liu, S.H. Chang, Eigenstates and fine structure of a hydrogenic impurity in a spherical quantum dot. Phys. Rev. B **58**, 1954–1961 (1998)
44. R.G. Toscano-Negrett, J.C. León-González, J.A. Vinasco, A.L. Morales, F. Koc, A.E. Kavruk, M. Sahin, M.E. Mora-Ramos, J. Sierra-Ortega, J.C. Martínez-Orozco, R.L. Retro, C.A. Duque, Optical properties in a ZnS/CdS/ZnS core/shell/shell spherical quantum dot : electric and magnetic field and donor impurity effects. Nanomaterials **13**, 550 (2023)
45. A.D. Yoffe, Semiconductor quantum dots and related systems: electronic, optical, luminescence and related properties of low dimensional systems. Adv. Phys. **50**, 1–208 (2001)
46. L. Shi, Z.W. Yan, Stark shift and photoionization cross section of on-center and off-center donor impurity in a core/shell ellipsoidal quantum dot. Physica E **98**, 111–117 (2018)
47. A. Cherni, N. Yahyaoui, N. Zeiri, P. Baser, M. Said, S. Saadaoui, Effect of dielectric mismatch on impurity binding energy, photoionization cross-section and stark shift of CdS/ZnSe core shell spherical quantumdots. Phys. Scr. **98**, 087001 (2023)
48. E.C. Niculescu, Interlevel transitions in core-shell nanodots with dielectric environment. Superlattices Microstruct. **51**, 814–824 (2012)
49. A. Ed-Dahmouny, M. Jaouane, R. Arraoui, A. Fakkahi, H.M. Althib, K. El-Bakkari, N. Zeiri, A. Sali, N. Es-Sbai, M.N. Murshed, C.A. Duque, Strain-induced modulations in nonlinear optical rectification of core/shell quantum dots within an MEH-PPV polymer matrix under electric field : a 3D finite-element modeling study. Discover Appl. Sci. **6**(11), 560 (2024)
50. P. Harrison, *Quantum Wells, Wires and Dots* (John Wiley & Sons Inc., New York, USA, 2000), pp.140–142
51. C. Kittel, Introduction to Solid State Physics, 8th Edn. ISBN #9780471415268
52. T. Ahamad, M.A. Majeed Khan, S. Kumar, M. Ahamed, M. Shahabuddin, A.N. Alhazaa, CdS quantum dots: growth, microstructural, optical and electrical characteristics. Appl. Phys. B **122**(6), 179 (2016)
53. J.P. Singh, S. Pal, A. Nag, Y.K. Sharma, Structural properties, energy interaction, and applications of CdS nanomaterial doped with rare-earth ions. Macromol. Symp. **413**, 2300169 (2024)
54. A.A. Adimado, Zinc Selenide, in book: Reference Module in Materials Science and Materials Engineering (2016)
55. M.T. Chowdhury, A. Zubair, H. Takeda, KMd.A. Hussain, Md.F. Islam, Optical and structural characterization of ZnSe thin film fabricated by thermal vapour deposition technique. AIMS J. **4**, 1095–1121 (2017)
56. M.M. Ivashchenko, A.S. Opanasyuk, I.P. Buryk, V.A. Lutsenko, A.V. Shevchenko, Optical properties of pure and eu doped ZnSe films deposited by CSVS technique. J. Nano-Electron. Phys. **9**, 01011 (2017)
57. V. Mehta, D. Shikha, J. Sharma, R.P. Chauhan, Temperature effect on properties of chemical induced nanocrystalline ZnSe thin films. J. Mater. Sci. Mater. Electron. **29**, 8801–8808 (2018)
58. M. Sundararajan, P. Sakthivel, A.C. Fernandez, Structural, optical and electrical properties of ZnO-ZnS nanocomposites prepared by simple hydrothermal method. J. Alloy. Compd. **768**, 553–562 (2018)
59. A. Rastogi, G. Pathak, A. Srivastava, J. Herman, R. Manohar, Cd1−X ZnXS/ZnS core/shell quantum dots in nematic liquid crystals to improve material parameter for better performance of liquid crystal based devices. J. Mol. Liq. **255**, 93–101 (2018)
60. S. Ummartyotin, Y. Infahsaeng, A comprehensive review on ZnS: from synthesis to an approach on solar cell. Renew. Sustain. Energy Rev. **55**, 17–24 (2016)
61. A. Chakrabarti, E. Alessandri, Syntheses, properties, and applications of ZnS-based nanomaterials. Appl. Nano **5**, 116–1421 (2024)
62. Y. Jiang et al., Homoepitaxial growth and lasing properties of ZnS nanowire and nanoribbon arrays. Adv. Mater. **18**, 1527–1532 (2006)

63. Z. Huang, X. Li, T. Liang, B. Ren, X. Zhang, Y. Zheng, Q. Zhang, Z. Fang, M. Wu, M. Zulfiqar, L. Jing, S. Qu, J. Chen, D.P. Gan, Smart mechanoluminescent phosphors: a review of zinc sulfide based materials for advanced mechano-optical applications. Responsive Mater. **2**, 1–52 (2024)
64. G. Hong, Seeing the sound. Science **369**, 638 (2020)
65. A. Rabanian, M. Neghabi, M. Zadsar, M. Jafari, Theoretical studies of energy states of CdSe/ZnS/CdSe and ZnS/CdSe/ZnS quantum dots with an impurity. Mater. Sci. Eng. B **274**, 115489 (2021)
66. N. Zeiri, A. Naifar, S. Abdi-Ben Nasrallah, M. Said, Dielectric environment effect on linear and nonlinear optical properties for CdS/ZnS core/shell quantum dots. Res. Phys. **15**, 102661 (2019)
67. K. Hasanirokh, A. Asgari, M. Mahdizadeh Rokhi, Theoretical study on nonlinear optical properties of CdS/ZnS spherical quantum dots. Optik **188**, 99–103 (2019)
68. S. Anchala, P. Purohit, K.C. Mathur, Linear and nonlinear intersubband optical properties of Si quantum dot embedded in oxide, nitride, and carbide matrix. J. Appl. Phys. **110**(11), 114320 (2011)
69. Z. Zeng, C.S. Garoufalis, A.F. Terzis, S. Baskoutas, Linear and nonlinear optical properties of ZnO/ZnS and ZnS/ZnO core shell quantum dots: Effects of shell thickness, impurity, and dielectric environment. J. Appl. Phys. **114**(2), 023510 (2013)
70. T.O. Cheche, V. Barna, Y.C. Chang, Analytical approach for type-II semiconductor spherical core–shell quantum dots heterostructures with wide band gaps. Superlattices Microstruct. **60**, 475–486 (2013)
71. L. Shi, Z.W. Yan, Electric field and shape effect on the linear and nonlinear optical properties of multi-shell ellipsoidal quantum dots. Superlattices Microstruct. **94**, 204–214 (2016)
72. L. Aderras, A. Bah, E. Feddi, L.M. BurileanuF, C.A.D. Dujardin, Stark-shift of impurity fundamental state in a lens shaped quantum dot. Phys. E **89**, 119–123 (2017)
73. S. M'zerd, M. El Haouari, M. Aghoutane, M. El-Yadri, E. Feddi, F. Dujardin, I. Zorkani, A. Jorio, M. Sadoqi, G. Long, Electric field effect on the photoionization cross section of a single dopant in a strained AlAs/GaAs spherical core/shell quantum dot. J. Appl. Phys. **124**(16) (2018)
74. A.L. Morales, N. Raigoza, E. Reyes-Gómez, J.M. Osorio-Guillén, C.A. Duque, Impurity-related polarizability and photoionization-cross section in GaAs-Ga $_{1-x}$Al$_x$As double quantum wells under electric fields and hydrostatic pressure. Superlattices Microstruct. **45**, 590–597 (2009)
75. H. Baghramyan, M. Barseghyan, A.A. Kirakosyan, R. Restrepo Arango, M. Mora-Ramos, C. Duque, Donor impurity-related linear and nonlinear optical absorptioncoefficients in GaAs=Ga1xAlxAs concentric double quantum rings: Effects of geometry, hydrostatic pressure, and aluminum concentration. J. Lumin. **145**, 676–683 (2014)
76. M.G. Barseghyan, M.E. Mora-Ramos, C.A. Duque, Hydrostatic pressure, impurity position and electric and magnetic field effects on the binding energy and photo-ionization cross section of a hydrogenic donor impurity in an InAs Pöschl-Teller quantum ring. Eur. Phys. J. B. **84**, 265–271 (2011)
77. K. Rahmani, Y. Chrafih, S. M'Zred, S. Janati, I. Zorkani, A. Jorio, A. Mmadi, Polarizability and binding energy of a shallow donor in spherical quantum dot-quantum well (QD-QW), in *IOP Conference Series: Journal of Physics: Conference Series*, vol. 984, pp. 012001(1) (2018)
78. A.L. Morales, N. Raigoza, E. Reyes-Gómez, J.M. Osorio-Guillén, C.A. Duque, Impurity-related polarizability and photoionization-cross section in GaAs-Ga$_{1-x}$Al$_x$As double quantum wells under electric fields and hydrostatic pressure. Superlattices Microstruct. **45**, 590–597 (2009)
79. A. Cherni, N. Yahyaoui, N. Zeiri, M. Said, S. Saadaoui, Simultaneous effect of the capped matrix and the geometric factors of CdS/ZnSe spheroidal quantum dots on the linear and nonlinear optical properties. Opt. Quant. Electron. **55**, 273–286 (2023)
80. A. Jbeli, N. Yahyaoui, N. Zeiri, M. Said, External electric field effect and impact of encapsulating matrix on optical and electronic properties within CdSe/ZnSe core/shell QDs. Phys. Scr. **96**, 125806 (2021)

81. K. Hasanirok, A. Radu, C.A. Duque, Donor impurity in CdS/ZnS spherical quantum dots under applied electric and magnetic fields. Nanomaterials **12**(22), 4014 (2022)
82. A. Cherni, N. Yahyaoui, N. Zeiri, P. Baser, M. Said and S. Saadaoui, Effect of dielectric mismatch on impurity binding energy, photoionization cross-section and stark shift of CdS/ZnSe core shell spherical quantumdots, Physica Scripta, vol. 98, pp. 087001 (2023)
83. N. Zeiri, A. Cherni, N. Yahyaoui, P. Baser, M. Said, S. Saadaoui, Photoionization cross-section and polarizability of impurity in CdS/ZnS core/shell quantum dots capped in a dielectric matrix. Solid State Commun. **368**, 115181 (2023)
84. N. Zeiri, A. Naifar, S. Abdi-Ben Nasrallah, M. Said, Third nonlinear optical susceptibility of CdS/ZnS core-shell spherical quantum dots for optoelectronic devices. Optik **176**, 162–167 (2019)
85. B. Kumar, B.K. Kaushik, Y.S. Negi, Perspectives and challenges for organic thin film transistors: Materials, devices, processes and applications. J. Mater. Sci. Mater. Electron. **25**, 1–30 (2014)

Donor Interaction Physics of Semiconductor QD Nano-Spheres

Sami Ortakaya

1 Introduction

Quantum nano-heterostructures have emerged as an important research topic in modern optoelectronic physics due to their ability to confine charge carriers at nanometer scale. This quantum confinement enables precise control over discrete energy levels, allowing for the design of optical transitions and improved efficiency in light-matter interactions [1]. Such control is essential for optimizing the performance of devices like quantum well lasers, light-emitting diodes (LEDs), and photodetectors, where electronic band structure engineering directly influences emission wavelengths, absorption spectra, and carrier dynamics [2]. Furthermore, the manipulation of charge transport mechanisms within these heterostructures enables the tuning of critical device parameters, including photodetector sensitivity, response time, and carrier transit times [3–5]. So that, the quantum nanostructures serve as fundamental building blocks in development of unique optoelectronic devices, offering enhanced performance, architecture, and functionality at quantum scale.

In recent years, the investigation of impurity-related effects in semiconductor quantum heterostructures has attracted considerable interest due to their significant influence on electronic and optical properties of nanoscale systems [6–13]. The donor atomic-interaction arises from impurity atoms embedded within the structure, which may release electrons that do not participate in bond formation. This non-bonding, loosely bound electrons introduce localized states within the heterostructure, resulting in modified energy eigenvalues that lie below those of free, unconfined carriers.

The presence of such impurity-induced bound states is inherently linked to Coulomb interactions, which alter not only the energy spectrum but also the spatial distribution of the carrier wavefunctions. Consequently, the wavefunction overlaps between allowed states, which are critical for determining optical transition probabil-

S. Ortakaya (✉)
Institute for Globally Distributed Open Research and Education (IGDORE), Gothenburg, Sweden
e-mail: sami.ortakaya@yahoo.com

R. En-nadir (ed.), *Quantum Systems at the Nanoscale*, Lecture Notes in Nanoscale Science and Technology 45, https://doi.org/10.1007/978-3-032-06645-9_2

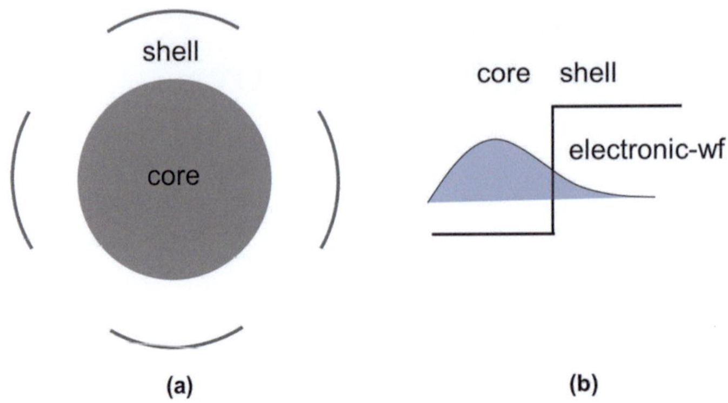

Fig. 1 Typical QD-sphere **a** core/shell structure and **b** conduction band profile for relevant heterostructure

ities, become significantly affected [14]. These alterations manifest in the absorption spectrum, oscillator strength, and other nonlinear optical responses of the material.

More intriguingly, in systems where two non-bonding electrons experience mutual Coulomb interaction without forming a conventional molecular bond, a helium-like configuration may emerge. This quasi-atomic state introduces additional complexity into the optical and electronic structure of the system, leading to rich physical behavior that necessitates careful theoretical and computational modeling [15–18].

In this chapter, we investigate the influence of such impurity configurations in quantum heterostructures, with a particular emphasis on the perturbation & self-consistent computing for energy eigenvalues. This analysis reveals how Coulomb-induced variations in spatial charge distribution lead to notable modifications in subband energies of the system. These results may provide new pathways for engineering optoelectronic responses in nanostructured devices.

2 Atomic Donor Arrangement in QDs

Let us consider a spherical QD that enables three-dimensional quantum confinement of electrons (Fig. 1). The transport of electrons located in the spherical semiconductor core with a lower band gap, or their interaction with optical waves, can be analyzed within the framework of quantum mechanics. This QD is composed of a typical semiconductor heterostructure, and the motion of electrons is confined in all spatial dimensions. In the optoelectronic analogy, confined electrons interact with the optical wave and are thereby excited to higher energy levels, allowing them to be transferred to the relevant contact via tunneling or transport at elevated energies. In fact, the active layer of an optical sensor system contains a large number of electrons within the conduction band, enabling transitions within the subband. In the absence of a bias voltage, it is possible to obtain transition probabilities that depend on the QD size in

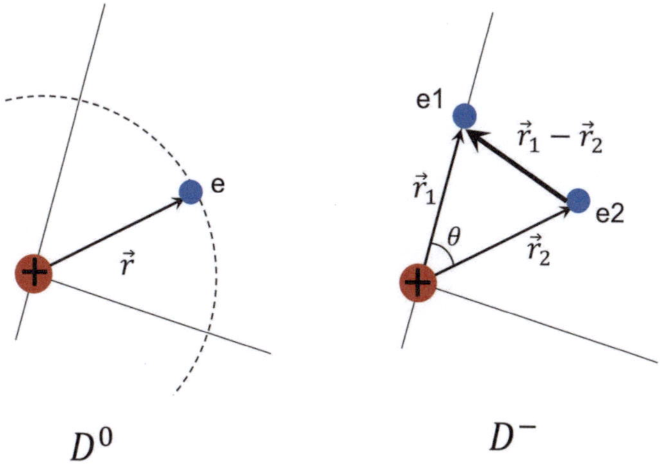

Fig. 2 Schematics of donor physics in the spherical regime

the active layer. Additionally, this configuration offers the advantage of comparing the quantum system with bulk materials [13]. It should be noted that, in the absence of heavily doped contacts and when only the active layer is considered, no potential curves leading to self-consistent solutions are formed.

In certain semiconductors, due to the relatively large band gap, excitation occurs predominantly through doping rather than thermal activation. In the case of donor doping, non-bonding electrons still experience Coulombic interaction with the heavy nucleus. When only one electron remains unbound, this condition is referred to as a "hydrogenic impurity," whereas if two electrons do not participate in the crystal lattice, the system exhibits a "helium-like impurity," which is distinct from a system with *two protons*. These electrons may contribute to transport, tunneling, and optical processes in a way that closely resembles the behavior of the system's pure quantum states. As can be seen in Fig. 2, on-center impurities have D^0 and D^- arrangements. The hydrogenic impurity is commonly denoted as D^0 , while in the case of two electrons within a single proton's influence, the donor is typically referred to as a D^- donor impurity.

3 Quantum Mechanical Aspects

As a conventional form, the electron in the D^0 model, known as the *hydrogenic impurity*, interacts with the heavy nucleus via Coulomb interaction. Within the effective mass approximation, the Hamiltonian of ith electron is given by

$$\hat{\mathcal{H}} = -\frac{\hbar^2}{2m^*}\nabla^2 - \frac{e^2}{4\pi\epsilon|\mathbf{r}|} + V_c(\mathbf{r}), \tag{1}$$

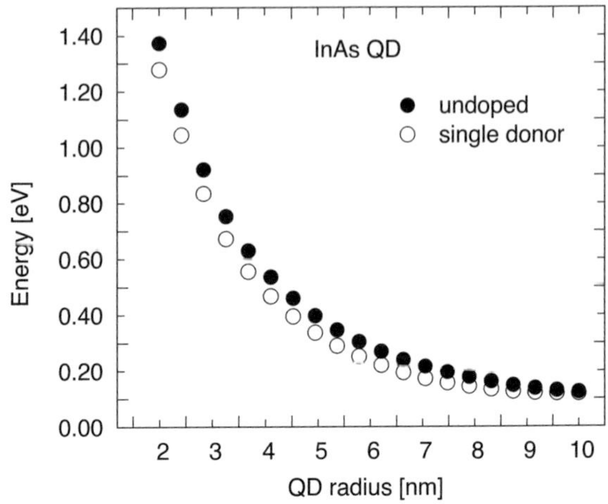

Fig. 3 D^0 interaction: Ground state $(n\ell m)$–(100) energy levels as a function of dot radius

where m^* denotes effective mass of electron, ϵ is the bulk dielectric constant of the material, $\mathbf{r}$ is integrated within the radial spherical coordinate, and V_c denotes the confinement potential energy. The corresponding energy and wavefunction are ε_i and $u(\mathbf{r}_i)$, respectively. For this well-studied model, existing research has provided data on quantum size effects [19], external fields [20], and optical coefficients [6, 21–23].

The decrease in energy values for the donor impurity can be showed in Fig. 3. Specifically, the ground state energy levels of a typical InAs quantum dot, modeled as an infinite potential well, are plotted as a function of the QD radius. Here, the effective mass is taken as $m^* = 0.027\, m_e$, and the dielectric constant corresponds to the bulk InAs value of $\epsilon = 15.08$ [4, 24, 25]. The reduction in energy values is expected, as the attractive Coulomb interaction shown in Eq. (1) lowers the resulting energy levels. Furthermore, by applying the established procedure for solving the Schrödinger equation, the results clearly demonstrate that the quantum size effect remains significant up to a QD radius of approximately 8.0 nm. Beyond this threshold, the binding energy approaches zero, indicating the suppression of confinement-induced interactions at larger sizes.

3.1 *Perturbation Theory of Nanocrystals*

In the D^- impurity, where "two electrons" are present, the repulsive interaction has been investigated through the Hartree self-consistent potential [26]. For the repulsive interaction, the perturbation contribution $\Delta\varepsilon$–arising from the individual energies ε_1 and ε_2 of the identical electrons corresponding to the Hamiltonian in Eq. (1)–can be written. The total energy of the D^- interaction, analogous to the perturbative Helium model, is calculated as

$$E_0 = \varepsilon_1(100) + \varepsilon_2(100) + \Delta E_0, \qquad (n\ell m) : (100) \tag{2}$$

Now, let us incorporate electron-electron interaction, D^- within perturbation theory and compute its contribution to the total energy. First, considering the ground state, it is well-known that when the spatial wavefunctions of two electrons are identical, the spin wavefunction must correspond to the antisymmetric vector $|00\rangle$ with total spin, $S = 0$. In this state, we can consider that the functions $\psi_a(\mathbf{r}_i)$ and $\psi_b(\mathbf{r}_i)$ (where $i = 1, 2$) exist. Hence, it can be observed that exchanging the indices results in a sign change in the spin wavefunction [27]. Therefore, the first-order contribution in the quantum state; $n\ell m$, can be given as follows:

$$\Delta E_0 = \frac{e^2}{4\pi\epsilon} \int\int \Psi^*(\mathbf{r}_1, \mathbf{r}_2) \frac{1}{|\mathbf{r}_1 - \mathbf{r}_2|} \Psi(\mathbf{r}_1, \mathbf{r}_2) d^3\mathbf{r}_1 d^3\mathbf{r}_2, \tag{3}$$

where

$$\Psi(\mathbf{r}_1, \mathbf{r}_2) = u_{n\ell m}(\mathbf{r}_1) u_{n\ell m}(\mathbf{r}_2)|00\rangle. \tag{4}$$

On the basis of electron-electron interaction without spin dependence, we can take into account only the spatial wave functions. In the ground state, both electrons occupy the same quantum state $(n\ell m) = (100)$, allowing for the singlet combination $|00\rangle$ (total spin, $S = 0$). Thus, the first-order perturbation of ground state becomes

$$\Delta E_0 = \frac{e^2}{4\pi\epsilon} \int\int |u_{100}(\mathbf{r}_1)|^2 \frac{1}{|\mathbf{r}_1 - \mathbf{r}_2|} |u_{100}(\mathbf{r}_2)|^2 \, d^3\mathbf{r}_1 \, d^3\mathbf{r}_2. \tag{5}$$

3.2 Excited States

In order to investigate optical absorption within the conduction-band structure of a QD nanosphere, it is necessary to determine the quantum numbers of the excited state. Therefore, according to the selection rule $\Delta\ell = \pm 1$, the total system in its excited state must have $\ell = 1$, and one of the electrons should be in the (100) state while the other is set in the (110) state. This is because if both electrons had the quantum number $\ell = 1$, the total spin would be zero ($S = 0$). However, a total spin of zero is only possible when both electrons have $\ell = 0$. Consequently, the wavefunction must be considered in the triplet form ($S = 1$) as shown in following form:

$$\Psi(\mathbf{r}_1, \mathbf{r}_2) = \frac{1}{\sqrt{2}} [u_{100}(\mathbf{r}_1) u_{110}(\mathbf{r}_2) - u_{100}(\mathbf{r}_2) u_{110}(\mathbf{r}_1)] \, |1, m\rangle, \quad m = 0, \pm 1, \tag{6}$$

where spin wave function has a well-known combination,

$$
\begin{aligned}
|0,0\rangle &= \frac{1}{\sqrt{2}}\left[\psi_a(1)\psi_b(2) - \psi_a(2)\psi_b(1)\right] \\
|1,1\rangle &= \psi_a(1)\psi_a(2) \\
|1,0\rangle &= \frac{1}{\sqrt{2}}\left[\psi_a(1)\psi_b(2) + \psi_a(2)\psi_b(1)\right] \\
|1,-1\rangle &= \psi_b(1)\psi_b(2).
\end{aligned}
\tag{7}
$$

At hence, the first-order perturbation of excited state energy reads

$$
E_e = \varepsilon_1(100) + \varepsilon_2(110) + \Delta E_e \tag{8}
$$

where

$$
\begin{aligned}
\Delta E_e = {} & \frac{e^2}{4\pi\epsilon}\int\!\!\int |u_{100}(\mathbf{r}_1)|^2 \frac{1}{|\mathbf{r}_1 - \mathbf{r}_2|} |u_{110}(\mathbf{r}_2)|^2 d^3\mathbf{r}_1\, d^3\mathbf{r}_2 \\
& - \frac{e^2}{4\pi\epsilon}\int\!\!\int u_{100}(\mathbf{r}_1)^* u_{110}(\mathbf{r}_2)^* \frac{1}{|\mathbf{r}_1 - \mathbf{r}_2|} u_{100}(\mathbf{r}_2) u_{110}(\mathbf{r}_1) d^3\mathbf{r}_1\, d^3\mathbf{r}_2.
\end{aligned}
\tag{9}
$$

3.3 Results for Ground State

It is possible to perform the calculation in Eqs. (5) and (9) without the need for a specific orientation of the $\mathbf{r}_1$ and $\mathbf{r}_2$ coordinates. The relative position of $\mathbf{r}_1$ and $\mathbf{r}_2$ remains the same regardless of the chosen axis. Therefore, we can choose the $\mathbf{r}_1$ coordinate along with $z = r\cos\theta$ axis. The angle between the vectors is then defined with respect to the z-axis, so we can write that

$$
\frac{1}{|\mathbf{r}_1 - \mathbf{r}_2|} = \frac{1}{\sqrt{r_1^2 + r_2^2 - r_1 r_2 \cos\theta}}. \tag{10}
$$

Here the volume elements are given with respect to the chosen axis. Fortunately, the single-particle wave functions are identical. Therefore, no restrictions apply within the integral domain. Since Eq. (5) relates to the perturbation correction, the energy shift can be written as

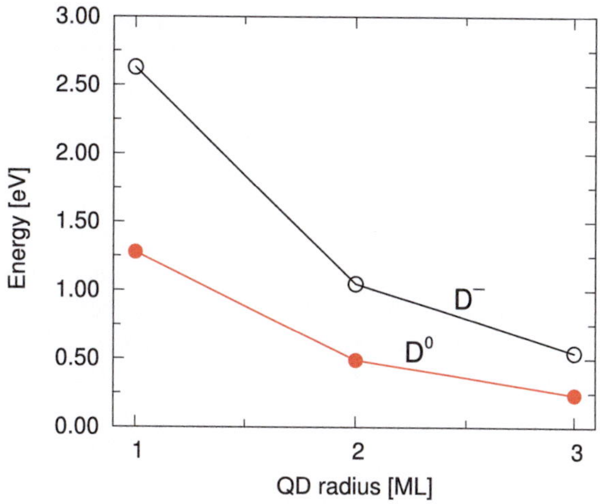

Fig. 4 Ground state (100) energy levels as a function of quantum dot radius

$$\Delta E_0 = \frac{2\pi e^2}{\epsilon} \int_0^{r_{\max}} dr_1\, r_1^2\, |u_{100}(r_1)|^2 \times \left[\frac{2}{r_1} \int_0^{r_1} dr_2\, r_2^2 |u_{100}(r_2)|^2 + 2 \int_{r_1}^{r_{\max}} dr_2\, r_2 |u_{100}(r_2)|^2 \right]. \tag{11}$$

(12)

The numerical results for the D^- donor impurity differs from that of the D^0. In this case, it is expected that the total energy arising from the repulsive interaction between the two electrons is greater than the sum of the single-particle energies. This expectation is confirmed in Fig. 4, as the value of ΔE_0 has been calculated to be 0.071 eV. Additionally, it is observed that the quantum confinement effect diminishes with increasing radius at monolayer scale (ML: 2 nm).

4 Hartree Self-Consistent Model for QD Nano-spheres

The perturbative approach becomes considerably complicated when determining the wave function of the excited states. The optical transition rate depends not only on the energy levels but also requires a quantitative evaluation of the overlap between the ground and excited states. To achieve this, one may solve the Schrödinger-Poisson coupling. The Schrödinger equation realted to the single-particle can be given in the form

$$\left[-\nabla_e \cdot \left(\frac{\hbar^2}{2\, m_e^*(r)} \nabla_e \right) + V_e(r) + V_c(r) \right] \Psi_k(\mathbf{r}) = E_k \Psi_k(\mathbf{r}), \tag{13}$$

where $m_e^*(r)$ is the electronic effective mass, $V_e(r)$ and $V_c(r)$ represent conduction-subband confinement potential and attractive Coulombic interaction, respectively. Considering the static dielectric constant ϵ, we can add central Coulomb interaction between electron (e1 and e2 are given in Fig. 2) and "on-center" donor nucleus in following relation

$$V_c(r) = -\frac{Ze^2}{4\pi\,\epsilon\, r}, \tag{14}$$

where Z is the number of heavy protons which takes the value, $Z = 2$ in Helium model. But now, we take that $Z = 1$ due to negative donor impurity (D^-) near QD center. The potential energy, including repulsive interaction and subband quantum confinement, is also given by

$$V_e(r) = U_e + q\,\Phi_H(r)\,, \tag{15}$$

where q represents the charge of single electron which are represented as $q = -e$, $\Phi_H(r)$ denotes Hartree energy for a given charge distribution, $N(\mathbf{r})$ which provides Poisson equation,

$$\nabla \cdot \big(\epsilon(r)\nabla\big)\Phi_H(r) = -q\,N(\mathbf{r})\,. \tag{16}$$

Here U_e also represents the subband confinement potential. We deal with spherical layers of the wave function

$$\Psi_{n\ell m}(r,\,\theta,\,\varphi) = R_{n\ell}(r)Y_{\ell m}(\theta,\,\varphi); \tag{17}$$

forthermore, this charge distribution satisfies that [28]

$$N(\mathbf{r}) = -\sum_j |\Psi_j(\mathbf{r})|^2, \tag{18}$$

with

$$\sum_j |\Psi_j(r,\,\theta,\,\varphi)|^2 = \frac{1}{4\pi}\sum_\ell 2(2\ell+1)\sum_n |R_{n,\ell}|^2 f_{\rm FD}(E_{n\ell}),\ n,\,\ell = 0,\,1,\,2,\,... \tag{19}$$

Here $f_{\rm FD}$ is the familiar Fermi-Dirac distribution at thermal energy, 0.026 eV. Considering the effective mass approach, the radial part of eigenfunctions are obtained from radial eigenvalue equation corresponds to $R(r)$. The radial eigenfunctions should behave as $R_{n\ell}$=0 (for $\ell\neq 1$) and R_{n0}=1 (for s-states) at center of nanosphere. In the presence of considered parameters, we solve radial part of Eq. (13) through trial energies under self-consistent calculations for convergence rule. If we use a spreadsheet has numerical data, required values can be easily obtained when self-consistent convergence occur via the smooth steps (h=0.01 nm). Namely, we construct the equations numerically, so we take relevant eigenvalues and numerical eigenfunctions. Inserting the two-electron interaction with Coulombic regime, we can also get the total energy of the form [14]

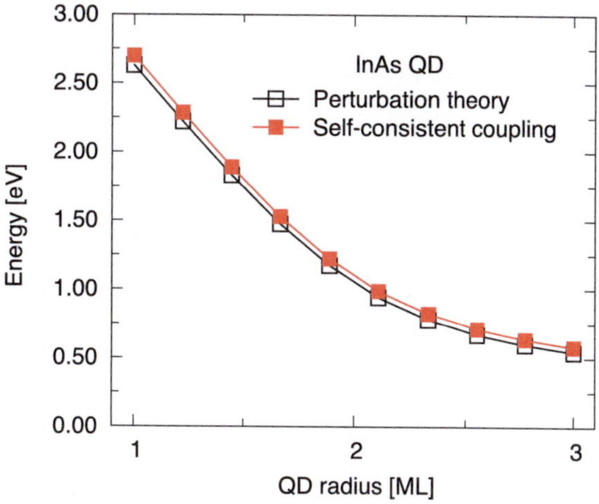

Fig. 5 Comparison of analytical and full numerical methods for the ground state energy levels of a negatively charged donor impurity (D^-) in the InAs QD nano-sphere

$$E_{\text{tot}} = \sum_{k=1}^{N} E_k - \int \Psi^* \Phi_H \Psi d^3\mathbf{r}. \tag{20}$$

The ground state energy E_0, obtained from Eq. (5) using the perturbation theory procedure, can be compared with the total ground state energy E_{tot} obtained from the solutions in Eq. (20). As illustrated in Fig. 5, the energy values obtained from both approaches exhibit good agreement. The consistency of these results, along with the applicability of the inverse-square law ($1/r^2$), confirms the reliability of the Hartree simulation for the D^- donor impurity mechanism.

5 Optical Aspects

Due to the difficulty of Eq. (9) in determining the excited states, $(n\ell m)$: (110), (120), (200), etc. for the two-electron D^- donor impurity, it is the greater way to work with the Hartree model rather than perturbation theory, especially in computing the wave function. We deal with the (110) excited-state because the relevant optical transition (100)→(110) is allowed by the $\Delta\ell = \pm 1$ rule. Therefore, we can obtain the energy level and the corresponding wave function of the (110) state by self-consistent computing.

The optical absorption in the conduction subbands gives the transition rate is given by [29]

$$w_{i\to f} = \frac{2\pi}{\hbar} |\langle i|f\rangle|^2 \cdot \delta(E_f - E_i - \hbar\omega), \tag{21}$$

where ω is the incident monochromatic optical frequency, and $|\langle i|f\rangle|$ represents the dipole matrix element between initial (i) and final (f) states. The transition rate

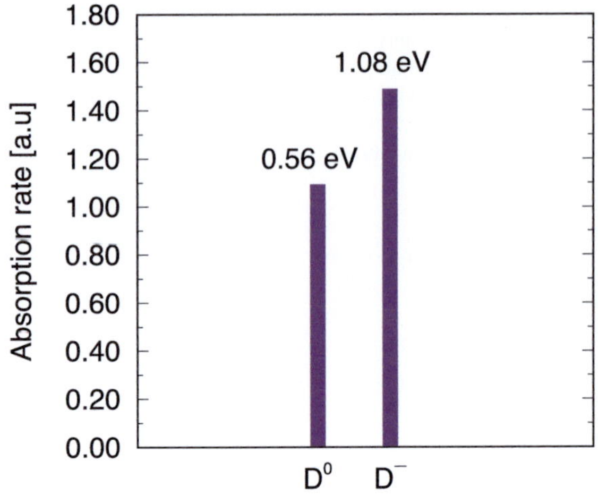

Fig. 6 Comparison of optical transition rate for neutral and negative donors in the InAs QD nano-sphere

linked to relevant absorption coefficient at the "resonance" peaks, so it is given by

$$w'_{i\to f} \propto |\langle i|e\cdot \mathbf{z}|f\rangle|^2 \cdot \frac{\hbar\omega\cdot\hbar\gamma}{(E_f - E_i - \hbar\omega)^2 + (\hbar\gamma)^2}, \tag{22}$$

where γ denotes relaxation time, $\gamma = 1/\tau$ related to scattering/phononic effectes. Considering transition energy, $E_f - E_i = \hbar\omega$, we obtain that peak-absorption rate becomes

$$W'_{i\to f} \propto (E_f - E_i)\cdot |\langle i|e\cdot \mathbf{z}|f\rangle|^2. \tag{23}$$

where we can consider that the polarization of incident optical wave is choosen in the direction, $z = r\cos\theta$. The absorption-related transition rate is presented in Fig. 6. For an InAs nanocrystal with a quantum dot radius of 3 nm, both neutral and negatively impurities are considered. Although the overlap values differ slightly, the negatively donor configuration yields a higher transition energy. Consequently, the absorption transition rate becomes larger. This implies that, for the D^- donor impurity, not only is the ground state energy affected, but also the energy difference between the first excited state (110) and the ground state (100) is greater compared to the D^0 configuration.

6 Conclusion

In this chapter, we have investigated the donor impurity effects of semiconductor quantum dot (QD) nanospheres by focusing on center-located donor nucleus. A detailed analysis was performed for both single- and two-electron configurations. The ground state energies obtained using the first-order perturbation were compared with those derived from Hartree self-consistent solutions. The good agreement between these approaches confirms the reliability of the computational framework and the validity of the approximations used, particularly in systems with confined geometries such as spherical QDs.

A significant outcome is the distinct behavior of optical absorption spectra when comparing neutral donors with negatively charged donors. While the spatial overlap between initial and final state wavefunctions may not show a substantial difference, the transition energies are notably higher for D^- states due to stronger Coulombic interactions and electronic repulsive effects. This difference results in enhanced absorption transition rates, which have important implications for designing optoelectronic devices operating in specific spectral regions.

Furthermore, the results have the critical role of donor configuration in determining the energy landscape and optical response of low-dimensional systems. As the size-dependent quantum confinement effects diminish with increasing nano-sphere radius, the influence of donor-induced perturbations becomes a dominant factor in tuning both electronic and optical properties. These results provide fundamental operations into tailoring QD-based architectures for high-performance applications in photodetectors, quantum image sensing, and nanoscale sensors. Overall, these results underscores the necessity of incorporating precise donor interaction models in theoretical and computational analyses of semiconductor QDs. The demonstrated sensitivity of both ground and excited state dynamics to impurity configurations reinforces the great importance of atomic physics in future nano-optoelectronic applicaitons.

Data Availability No Data associated in the manuscript.

Conflict of Interest The author declares that he has no confict of interest.

References

1. D. Indjin, *Optical Properties of Quantum Wells* (John Wiley & Sons, Ltd, 2016), pp. 379–406. https://doi.org/10.1002/9781118923337.ch11. URL https://onlinelibrary.wiley.com/doi/abs/10.1002/9781118923337.ch11
2. S. Chuang, *Physics of Photonic Devices*. Wiley Series in Pure and Applied Optics (Wiley, 2009). URL https://books.google.com.tr/books?id=x5Cd__PDf1kC
3. J. Singh, *Semiconductor Optoelectronics: Physics and Technology*. Electronics and VLSI circuits (McGraw-Hill, 1995). URL https://books.google.com.tr/books?id=IcYrnQAACAAJ
4. G. Zhou, P. Runge, IEEE J. Quantum Electron. **50**(4), 220 (2014). https://doi.org/10.1109/JQE.2014.2305015
5. H. Boerma, T. Kieckhefel, T.T. Tran, P. Runge, M. Schell, Polarization-independent photodetector with integrated optical preamplifier and 60 ghz 3 db bandwidth, in *Optical Fiber Communication Conference (OFC) 2024* (Optica Publishing Group, 2024), p. Tu3D.6. https://doi.org/10.1364/OFC.2024.Tu3D.6
6. S. Ortakaya, M. Kirak, A. Guldeste, J. Nonlinear Opt. Phys. Mater. **26**(03), 1750035 (2017). https://doi.org/10.1142/S0218863517500357

7. G.A. Mantashian et al., Atoms **9**(4), 75 (2021). https://doi.org/10.3390/atoms9040075
8. L. Pulgar-Velásquez et al., Nanomaterials **11**, 2832 (2021). https://doi.org/10.3390/nano11112832
9. K. Hasanirokh, A. Radu, C.A. Duque, Nanomaterials **12**(22), 4014 (2022). https://doi.org/10.3390/nano12224014
10. R. Khordad, A. Ghanbari, K. Abbasi, A. Ghaffaripour, J. Comput. Electron. **22**(1), 260 (2023). https://doi.org/10.1007/s10825-022-01985-8
11. A. Kachu, A. Boda, Phys. E **155**, 115826 (2024). https://doi.org/10.1016/j.physe.2023.115826
12. S. Ortakaya, C.A. Duque, Appl. Phys. A **130**, 381 (2024). https://doi.org/10.1007/s00339-024-07509-4
13. S. Ortakaya, C.A. Duque, Phys. Scr. **99**, 045968 (2024). https://doi.org/10.1088/1402-4896/ad328a
14. R. Aydin, H. Tas, M. Sahin, Superlattices Microstruct. **86**, 292 (2015)
15. D.V. Melnikov, J.P. Leburton, in *Physical Models for Quantum Dots* (Jenny Stanford Publishing, 2021), pp. 237–261
16. S. Li, L. Shi, Z.W. Yan, Mod. Phys. Lett. B **34**(14), 2050153 (2020)
17. E. Al, Indian J. Phys. pp 1–10 (2024)
18. K. Bayrak, D. Kaya, O. Bayrak, Eur. Phys. J. Plus **139**(8), 719 (2024)
19. N. Ilnytskyi, et al., Phys. Chem. Solid State **25**, 98 (2024). https://doi.org/10.15330/pcss.25.1.98-103
20. N. Amin, A.J. Peter, Phys. B: Condensed Matter. **631**, 413693 (2022). https://doi.org/10.1016/j.physb.2022.413693
21. V. Holovatsky, M. Chubrei, V. Ivanko, Optical absorption in core-shell quantum antidot with donor impurity under applied magnetic field, in *2021 IEEE 11th International Conference Nanomaterials: Applications & Properties (NAP)* (2021), pp. 1–5. https://doi.org/10.1109/NAP51885.2021.9568536
22. N. Zeiri et al., Solid State Commun. **368**, 115181 (2023). https://doi.org/10.1016/j.ssc.2023.115181
23. H. Xie, X. Zhou, Y. Xing, Micro. Nanostruct. **195**, 207948 (2024). https://doi.org/10.1016/j.micrna.2024.207948
24. S. Ortakaya, Res. Opt. **13**, 100512 (2023)
25. A. Bouazra, S. Mnasri, S. Abdi-Ben Nasrallah, M. Said, Comput. Phys. Commun. **185**(4), 1290 (2014). https://doi.org/10.1016/j.cpc.2014.01.003
26. F. Koç, A.E. Kavruk, M. Sahin, Phys. E: Low-dimensional Syst. Nanostruct. **145**, 115479 (2023). https://doi.org/10.1016/j.physe.2022.115479
27. D.J. Griffiths, D.F. Schroeter, *Introduction to Quantum Mechanics*. (Cambridge University Press, 2018)
28. J. Sée, P. Dollfus, S. Galdin, J. Appl. Phys. **92**(6), 3141 (2002) https://doi.org/10.1063/1.1499524
29. A.M. Fox, *Optical Properties of Solids*. Oxford Master Series in Condensed Matter Physics (Oxford University Press, 2001)

Normalized Exciton Binding Energy in Cylindrical Nanostructures: Influences of Quantum Confinement and LO Phonon Contributions

M. Hbibi, S. Chouef, M. El Hadi, R. Boussetta, A. El Moussaouy, E. H. Ouacha, N. El Biaze, O. Mommadi, and C. A. Duque

1 Introduction

The study of nanostructures has become a cornerstone in various scientific disciplines, including physics, chemistry, material science, and biology [1, 2]. Modern fabrication technologies enable accurate control over structures at the nanometer scale, paving the way for rapid advances in nanotechnology. Among these, semiconductor nanostructures stand out due to their tunable properties, achievable through nanoscale engineering of crystalline materials. Prior investigations have extensively analyzed the behavior of quasi-particles such as electron-impurity, excitons, trions, and biexcitons, along with their coupling to phonons, particularly in systems like

M. Hbibi · S. Chouef · M. El Hadi · A. El Moussaouy
OAPM Group, Laboratory of Materials, Waves, Energy and Environment, Department of Physics, Faculty of Sciences, University Mohamed I, Oujda, Morocco

R. Boussetta
LRDSI, Department of Physics, Al-Hoceima Faculty of Science and Technology, Abdelmalek Essaâdi University, Tétouan, Morocco

A. El Moussaouy
Laboratory of Innovation in Science, Technologie and Education, CRMEF, Oujda, Morocco

E. H. Ouacha
Laboratory of Metrology and Information Processing, Faculty of Sciences, Ibnou Zohr University, Agadir, Morocco

E. H. Ouacha · N. El Biaze · O. Mommadi (✉)
Research Laboratory in Sciences and Techniques, ESEFA, Ibnou Zohr University, Agadir, Morocco
e-mail: o.mommadi@uiz.ac.ma

N. El Biaze
Materials and Renewable Energy Laboratory, Ibnou Zohr University, Agadir, Morocco

C. A. Duque
Grupo de Materia Condensada-UdeA, Instituto de Física, Facultad de Ciencias Exactas y Naturales, Universidad de Antioquia UdeA, Medellín, Colombia

R. En-nadir (ed.), *Quantum Systems at the Nanoscale*, Lecture Notes in Nanoscale Science and Technology 45, https://doi.org/10.1007/978-3-032-06645-9_3

graphene and quantum dots (QDs) [3–15]. While our understanding of quantum correlated states remains incomplete, these entities are crucial for emerging technologies in optoelectronics, quantum information processing, energy storage, and photovoltaics [15–17]. Thus, exploring how quantum confinement influences excitonic behavior is fundamental for developing efficient nanostructured devices.

Cylindrical nanostructures offer rich possibilities for adjusting their electronic and optical characteristics by varying dimensions such as radius and thickness. These modifications impact properties such as energy quantization, optical response, and structural stability. By increasing cross-sectional size, systems can transition from zero-dimensional quantum dots to quasi-one- or two-dimensional forms like wires and wells, thereby affecting exciton dynamics [18, 19]. The interplay between geometric confinement, dielectric mismatch, and external stimuli provides pathways to tailor materials for specific applications, such as solar energy harvesting and light emission [20–27]. Nevertheless, the role of exciton-phonon coupling remains underexplored in many such investigations. Experimental and theoretical works have shown that phonons, particularly longitudinal optical phonons (LOPs), influence key optoelectronic features, including excitonic recombination lifetimes and spectral characteristics. Accurate modeling of these interactions enhances device performance and reliability.

With growing focus on cylindrical QDs in recent literature, it becomes evident that the normalized binding energy of excitons in such systems—especially under polaronic corrections from LOP modes has not been fully characterized. Building on our previous studies [26, 28], we aim to elucidate the influence of phonon modes on excitonic states, highlighting their relevance to device engineering. The confinement potential, radius, height, and aluminum concentration in the surrounding barrier material collectively determine the electronic structure. Despite their significance, these parameters' combined effects on exciton-LOP interactions in cylindrical QDs with finite potential barriers have not been sufficiently addressed. This work aims to bridge this gap by analyzing how the spatial confinement and material composition modulate the normalized binding energy of excitons.

Our theoretical model assesses the influence of LOP interactions on excitonic ground state energies in cylindrical nanostructures, where the variable aluminum content in the barrier region serves as a control parameter for the confinement potential. These insights are expected to contribute to the rational design of optoelectronic components, such as quantum dot-based LEDs.

The structure of the paper is as follows: Sect. 2 describes the theoretical approach for excitonic modeling; Sect. 3 presents results concerning the normalized binding energy; and Sect. 4 summarizes the conclusions drawn from our findings.

2 Theoretical Framework

We consider a cylindrical quantum dot (QD) structure composed of GaAs, characterized by a radius R and height H, and embedded in a surrounding barrier made of

$Ga_{1-x}Al_xAs$. The system is described using the effective mass approximation, under which the total Hamiltonian governing the excitonic behavior, including phonon contributions, is expressed as:

$$H_{Sys} = H_{ex} + H_{LOP} + H_{ex-LOP}, \tag{1}$$

where H_{ex} represents the Hamiltonian of a non-interacting exciton, H_{LOP} corresponds to the longitudinal optical phonon (LOP) term, and H_{ex-LOP} accounts for the coupling between the exciton and phonon field [26].

To evaluate the ground state excitonic binding energy in the presence of phonons, we use a variational wave function of the form [18]:

$$\psi_{ex}(\mathbf{r_e}, \mathbf{r_h}) = \psi_e(\mathbf{r_e})\psi_h(\mathbf{r_h})\psi_{eh}(\mathbf{r}_{eh}), \tag{2}$$

where the correlation between the electron and hole is modeled as:

$$\psi_{eh}(\mathbf{r}_{eh}) = \exp\left(-\alpha\rho_{eh}\right)\exp\left(-\gamma(z_e - z_h)^2\right), \tag{3}$$

with variational parameters α and γ. The in-plane relative distance is given by $\boldsymbol{\rho}_{\mathbf{eh}} = \boldsymbol{\rho}_{\mathbf{e}} - \boldsymbol{\rho}_{\mathbf{h}}$.

The confinement potential is assumed to be finite and defined separately for radial and axial components. The radial (lateral) square potential is modeled as [26]:

$$V_i(\rho_i) = \begin{cases} 0, & if \;\; 0 < \rho_i < R \\ V_i(x), & if \;\; \rho_i > R \end{cases}, \tag{4}$$

and the parabolic potential along the axial direction as:

$$V_i(z_i) = \begin{cases} 0, & if \;\; -H/2 < z_i < H/2 \\ 4 \times \frac{Z_i^2}{H^2} V_i(x), & otherwise \end{cases}. \tag{5}$$

Here, the potential barrier for electrons and holes depends on the aluminum concentration x through:

$$V_e(x) = 0.658 \times \left(1.155x + 0.37x^2\right), \tag{6}$$

$$V_h(x) = 0.342 \times \left(1.155x + 0.37x^2\right). \tag{7}$$

The phonon-exciton binding energy is calculated as:

$$E_B^{ph} = E_e^{ph} + E_h^{ph} - E_{ex}^{ph}, \tag{8}$$

where E_e^{ph} and E_h^{ph} are the energies of the electron and hole, respectively, and E_{ex}^{ph} is the total ground state excitonic energy in the presence of phonons, obtained by minimizing the following functional:

$$E_{ex}^{ph} = \min_{\alpha,\gamma} \langle \psi_{ex} \mid H_{Sys} \mid \psi_{ex} \rangle . \quad (9)$$

To assess confinement effects, we introduce the normalized exciton binding energy, defined as [29, 30]:

$$NE_B = \frac{E_B^{ph}}{E_{ex}^{ph}} \quad (10)$$

3 Numerical Results and Discussion

The theoretical framework outlined earlier is now applied to investigate the normalized binding energy of exciton coupled with LOP modes. The composition of the barrier material typically includes the aluminum concentration ranging between 0.1 and 0.4%, to keep the direct bandgap. The key physical parameters considered in this investigation include: $m_e^* = 0.067\, m_0$, $m_h^* = 0.087\, m_0$, $\varepsilon_0(\omega) = 12.83$, $\varepsilon_\infty(\omega) = 10.9$, LO-phonon energy: $\hbar\omega_{LO} = 36.25$ meV, $a_{ex}^* = 18$ nm, $R_{ex}^* = 3.0$ meV and the exciton-phonon coupling $\alpha_F = 0.0681$.

Figure 1 presents the variation of the normalized exciton binding energy NE_B as a function of the cylindrical radius R, for three different nanostructure thicknesses: $H = 0.5a_{ex}^*$, $1a_{ex}^*$, and $2a_{ex}^*$, with the aluminum concentration in the barrier material fixed at $Al_x = 0.3$. The exciton is considered to be confined within the cylindrical structure and interacts with LOP. The inclusion of LOP interactions introduces a significant correction to the exciton energy, particularly in the strong confinement regime (small R and H), where the interaction between the exciton and the lattice vibrations becomes more pronounced [26, 28]. The normalized binding energy is calculated using an approach analogous to Eq. (10), which includes both quantum confinement and impurity effects. The figure shows a clear monotonic decrease in the normalized binding energy NE_B as the radius R increases for all values of H in the presence of LOP. This polaronic effect contributes to the increased binding energy in narrow nanostructures, emphasizing the importance of phonon coupling in realistic systems. This behavior is a direct consequence of the quantum confinement effect: smaller radii enhance the spatial confinement of the electron-hole pair, leading to a stronger Coulomb attraction and thus a higher exciton binding energy [26]. As the radius increases, the confinement weakens and the binding energy asymptotically decreases. Additionally, for a given radius, the exciton binding energy is observed to be higher for smaller values of thickness H. The curve corresponding to $H = 0.5a_{ex}^*$ exhibits the highest binding energies, while that for $H = 2a_{ex}^*$ shows the lowest. This indicates that reducing the nanostructure's height enhances the confinement in the

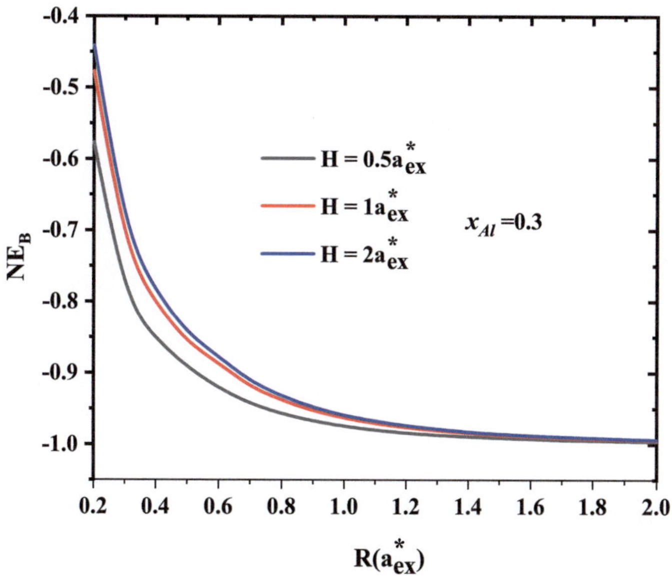

Fig. 1 Normalized exciton binding energy NE_B as a function of the cylindrical radius R for different nanostructure thicknesses: $H = 0.5a^*_{ex}$ (black), $H = 1a^*_{ex}$ (red), and $H = 2a^*_{ex}$ (blue), with the aluminum concentration in the barrier material fixed at $Al_x = 0.3$. The calculations consider a confined exciton interacting with LOP

axial direction, thereby further increasing the Coulomb interaction and the resulting excitonic binding energy.

Figure 2 illustrates the dependence of the normalized exciton binding energy NE_B on the normalized nanostructure thickness H/a^*_{ex}, for three different fixed values of the cylindrical radius: $R = 0.5a^*_{ex}$, $1a^*_{ex}$, and $2a^*_{ex}$. The calculations assume a confined exciton interacting with LOP, and an aluminum concentration in the barrier material of $Al_x = 0.3$. As shown in the figure, for all considered radii, the normalized exciton binding energy increases (i.e., becomes less negative) as the thickness H increases. This behavior is characteristic of reduced confinement in the axial direction: as the structure becomes thicker, the exciton behavior has less spatial restriction along the z-axis, which leads to a weakening of the Coulomb interaction and hence a reduction in the binding energy. This behavior is due to the parabolic confinement potential in the z-axis. For smaller radii ($R = 0.5a^*_{ex}$, black curve), the binding energy exhibits a more pronounced increase with thickness, particularly in the range $H < 0.6a^*_{ex}$. This sensitivity indicates strong quantum confinement in both radial and axial directions, which significantly enhances the exciton-phonon interaction and the Coulombic attraction at small dimensions. A slight non-monotonic feature is observed around $H \approx 0.4a^*_{ex}$, which may be attributed to the interplay between phonon coupling and the transition from strong to intermediate confinement regimes. However, as the thickness increases and passes the value $H \approx 0.3a^*_{ex}$, the axial confinement weakens, and the radial dimension begins to play a more dominant role

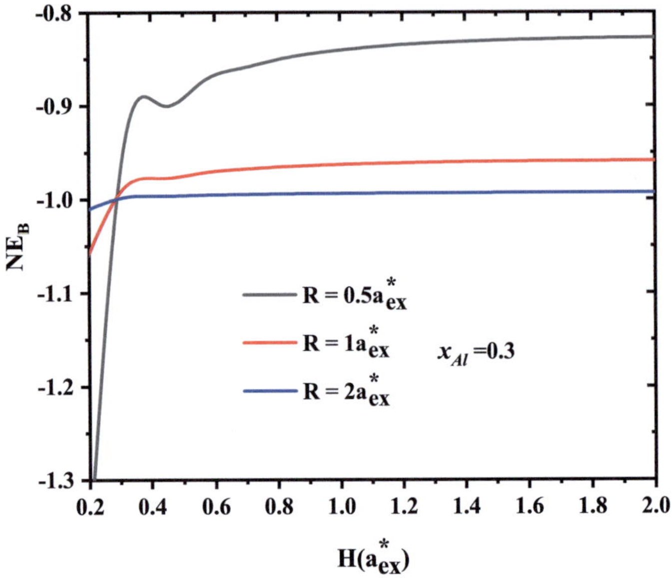

Fig. 2 Normalized exciton binding energy NE_B as a function of nanostructure thickness H, for different cylindrical radii $R = 0.5a^*_{ex}$ (black), $R = 1a^*_{ex}$ (red), and $R = 2a^*_{ex}$ (blue), at a fixed aluminum concentration $Al_x = 0.3$. The calculations include exciton-LOP interactions

in determining the exciton's spatial distribution and energy. At this crossover point, the influence of R becomes negligible, and the normalized exciton binding energy becomes effectively independent of the radius. This intersection can be interpreted as a critical point, marking a transition from a regime dominated by 3D-quantum confinement to a quasi-1D confinement regime, in which the exciton is primarily confined in the radial direction, while it becomes free along the axial direction. Beyond this point, the curves corresponding to larger radii exhibit less negative binding energies, indicating weaker overall confinement, and the order of the curves reflects this change in confinement character. In contrast, for larger radii ($R = 1a^*_{ex}$ and $2a^*_{ex}$, red and blue curves, respectively), the variation in normalized binding energy with thickness is less significant. These curves tend to saturate as H increases, indicating that the exciton becomes quasi-2D, with the confinement primarily governed by the radial dimension. At $R = 2a^*_{ex}$, the system exhibits a nearly flat response of NE_B to changes in H, suggesting that thickness plays a minor role in the excitonic behavior when the radial confinement is weak. The overall results demonstrate that the quantum confinement along the axial direction has a strong effect only in highly confined structures (small R), and that LOP interactions intensify this effect in the strong confinement regime. These findings reaffirm the need to consider both radial and axial dimensions when designing nanostructures for targeted excitonic and optoelectronic properties.

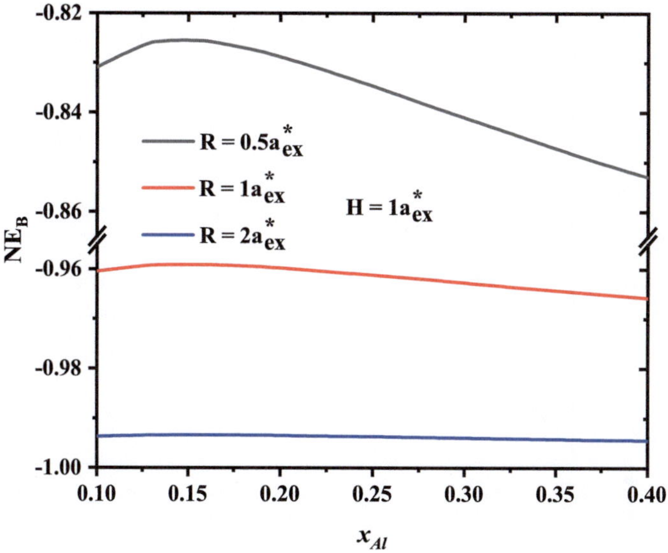

Fig. 3 Normalized exciton binding energy NE_B as a function of aluminum concentration x_{Al} in the barrier material, for different cylinder radii: $R = 0.5a^*_{ex}$ (black), $R = 1a^*_{ex}$ (red), and $R = 2a^*_{ex}$ (blue), with the height fixed at $H = 1a^*_{ex}$

Figure 3 presents the variation of the normalized exciton binding energy NE_B as a function of the aluminum concentration Al_x in the barrier material, for three different radii of the cylindrical nanostructure: $R = 0.5a^*_{ex}$, $1a^*_{ex}$, and $2a^*_{ex}$, while the thickness is fixed at $H = 1a^*_{ex}$. As the Al concentration increases, the barrier potential becomes higher, which strengthens the quantum confinement of the electron-hole pair within the nanostructure. This confinement effect leads to an increase in the exciton binding energy E_B. However, it simultaneously increases the total minimized excitonic energy E_{min} due to greater kinetic and potential energy contributions from the tighter confinement. Because NE_B is the ratio E_B/E_{min}, its evolution reflects the competition between the increasing Coulomb interaction (which enhances E_B) and the rising confinement energy (which boosts E_{min}). This interplay explains why the curves in the figure exhibit a non-monotonic or weakly decreasing behavior with increasing Al_x, especially at smaller radii where confinement effects are strongest. For small radius $R = 0.5a^*_{ex}$, a peak appears near $Al_x \approx 0.15$, indicating an optimal balance between Coulomb attraction and confinement. At larger radii ($R = 1a^*_{ex}$, $R = 2a^*_{ex}$), the variation becomes more gradual, as the exciton becomes less sensitive to the barrier composition due to reduced confinement. These results highlight that normalized binding energy, when defined relative to the total minimized excitonic energy, takes concessions between interaction and confinement energies, and should not be interpreted in the same way as absolute or Rydberg-normalized binding energies.

Figure 4 presents the evolution of the normalized exciton binding energy NE_B as a function of the aluminum concentration Al_x of the barrier material. The radius is fixed

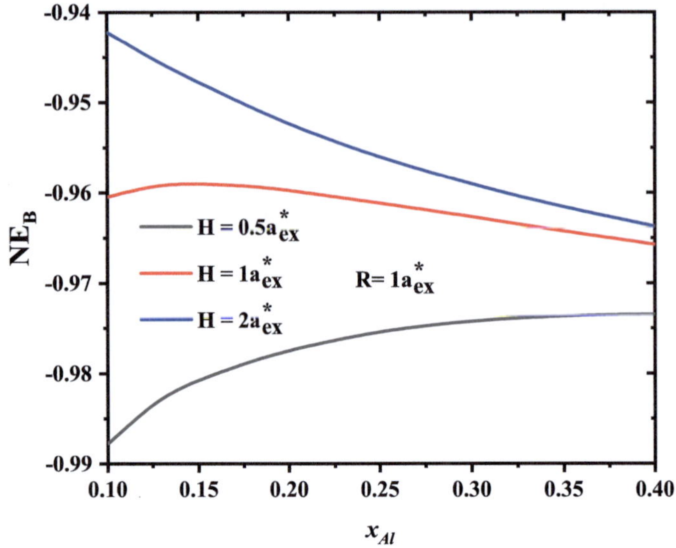

Fig. 4 Normalized exciton binding energy NE_B as a function of aluminum concentration Al_x in the barrier material for different cylinder thicknesses $H = 0.5, 1, 2a^*_{ex}$, with fixed radius $R = 1a^*_{ex}$

at $R = 1a^*_{ex}$, while the nanostructure thickness varies across three values: $H = 0.5, 1$, and $2a^*_{ex}$. For small thickness $H = 0.5a^*_{ex}$, the exciton experiences strong axial confinement, and increasing Al_x—which raises the barrier height leads to stronger localization and enhanced Coulomb interaction, resulting in a monotonic increase of NE_B. In contrast, at large thickness $H = 2a^*_{ex}$, the exciton is more spatially extended along the axial direction. In this regime, although the binding energy increases with Al_x, the minimized total excitonic energy increases faster due to stronger kinetic confinement, leading to a net decrease in NE_B. Interestingly, for the intermediate case $H = 1a^*_{ex}$, a weak maximum is observed around $Al_x \approx 0.2$, reflecting a crossover where the relative influence of Coulomb binding and confinement energy shifts. This non-monotonic behavior highlights the delicate balance between quantum confinement effects and material-induced potential modulation. Overall, these results reveal that the normalized exciton binding energy does not follow a simple trend with increasing aluminum concentration. Instead, its behavior is governed by the competition between charge carrier localization (enhancing binding) [29, 30] and the increase in total energy due to stronger confinement a dependence that is strongly modulated by the nanostructure's geometry.

The insights gained from the analysis of normalized exciton binding energy in cylindrical nanostructures have significant implications for the design and optimization of optoelectronic devices such as quantum dot lasers, photodetectors, and light-emitting diodes. By tuning geometric parameters (radius and thickness) and material composition (Al content), it becomes possible to precisely engineer excitonic properties for targeted performance. In particular, controlling the confinement regime enables enhanced exciton stability and binding, which is critical for improving efficiency in low-dimensional systems where strong Coulomb interactions dominate. These results provide a theoretical foundation for tailoring nanostructure-based components in advanced photonic and quantum information technologies.

4 Conclusions

In this work, we have investigated the behavior of the normalized exciton binding energy in cylindrical nanostructures, taking into account the influence of LOP interactions, geometric confinement, and barrier composition. By employing the effective mass approximation combined with a variational approach, we have demonstrated how the excitonic properties are strongly influenced by the nanostructure dimensions (radius and thickness) and the aluminum concentration in the surrounding barrier material. The results reveal a rich interplay between quantum confinement and phonon coupling, leading to non-monotonic trends in the normalized binding energy depending on the confinement regime. Notably, crossover points and curve reordering indicate dimensional transitions in exciton behavior. These findings provide valuable insights for the design and engineering of nanostructured optoelectronic devices, where precise control of excitonic interactions is essential for enhancing device performance.

Acknowledgements CAD is grateful to the Colombian Agencies: CODI-Universidad de Antioquia (Estrategia de Sostenibilidad de la Universidad de Antioquia and projects “Complejos excitónicos y propiedades de transporte en sistemas nanométricos de semiconductores con simetría axial basados en GaAs e InAs”, and “Nanoestructuras semiconductoras con simetría axial basadas en InAs y GaAs para aplicaciones en electrónica ultra e híper rápida”) and Facultad de Ciencias Exactas y Naturales-Universidad de Antioquia (CAD exclusive dedication project 2025–2026).

References

1. H. Rao et al., J. Mater. Chem. A **8**, 10233–10241 (2020)
2. D. Vasudevan et al., J. Alloys Compd. **636**, 395–404 (2015)
3. R. Boussetta et al., Phys. B **665**, 415009 (2023)
4. M. Chnafi et al., Optik **310**, 171881 (2024)
5. S. Chouef et al., Solid State Phenom. **335**, 31–41 (2022)
6. S. Ortakaya, Results Opt. **13**, 100512 (2023)
7. L. Belamkadem et al., Phys. E **129**, 114642 (2021)
8. R. Boussetta et al., Phys. B **680**, 415527 (2024)
9. L. Belamkadem et al., J. Nanomater. **12**, 966 (2022)
10. M. Chnafi et al., Superlattices Microstruct. **159**, 107052 (2021)
11. S. Chouef et al., Solid State Phenom. **350**, 137–143 (2023)
12. S.N. Gupta et al., Nat. Commun. **12**, 1310 (2021)
13. M. Chnafi et al., Phys. B Condens. Matter **594**, 412333 (2020)
14. M. Hbibi et al., Optik **324**, 172229 (2025)

15. S. Mueller et al., ACS Nano **15**, 4647–4657 (2021)
16. N.P. De Leon et al., Science **372** (2021)
17. R.D. Septianto et al., NPG Asia Mater. **12** (2022)
18. M. Hbibi et al., Sci. Rep. **12** (2022)
19. M. Hbibi et al., Sol. Energy **15**, 113620 (2025)
20. M. Chizaki et al., Mater. Sci. Semicond. Process. **148**, 106794 (2022)
21. D.I. Ilić et al., Acta Phys. Pol. **136** (2019)
22. Y. Liu et al., Mod. Phys. Lett. B **33**, 1950068 (2019)
23. A. Talbi et al., Appl. Phys. A **127** (2021)
24. A. Ed-Dahmouny et al., Results Phys. **74**, 108281 (2025)
25. M. Jaouane et al., Eur. Phys. J. Plus **138**, 1–10 (2023)
26. O. Mommadi et al., Phys. E **118**, 113903 (2022)
27. T.D. Lee et al., Phys. Rev. B **90** (1953)
28. O. Mommadi et al., Mater. Sci. Semicond. Process. **168**, 107805 (2023)
29. C. Dane et al., Phys. E **44**, 186–189 (2011)
30. H. Akbas et al., Superlattices Microstruct. **94**, 131–137 (2016)

The Nonlinear Optical Rectification Coefficient in a MSQD: Effects of External Perturbation and the Kratzer Confinement Potential

A. Fakkahi, M. Jaouane, H. Azmi, A. Ed-Dahmouny, A. Sali, R. Arraoui, K. El-Bakkari, O. Ben Hammou, S. El Otmani, J. El-Hamouchi, M. Jaafar, A. Mazouz, and H. El-Ghazi

1 Introduction

Nonlinear optics, which studies the effects and interactions of intense coherent light radiation with matter, has been a rapidly growing and fast-expanding field in recent years. Nonlinear optical effects in nanostructures, which have a great number of applications such as far-infrared laser amplifiers [1] and photodetectors [2], are higher than that in bulk materials as a result of the presence of potent quantum confinement. Nonlinear optical rectification (*NOR*), which is an important second-order nonlinear optical process, was first discovered by Bass et al. [3] in 1962. After experimental observations, a wide variety of theoretical investigations are done about *NOR* for the different shapes of quantum dots, such as spherical [4], cylindrical [5], cubic [6], lens [7], double cone like [8], etc. In recent years, several theoretical studies have been conducted to investigate the optoelectronic characteristics of low-dimensional semiconductor nanostructures such as the quantum well (QW), quantum well wire (QWW), and quantum dot (QD) [9–15], because of their potential applications in different areas such as quantum optics, optoelectronics, and quantum computation

A. Fakkahi (✉) · M. Jaouane · H. Azmi · A. Sali · R. Arraoui · K. El-Bakkari · O. B. Hammou · S. El Otmani · J. El-Hamouchi · M. Jaafar · A. Mazouz
Laboratory of Solid-State Physics, Department of Physics, Faculty of Science Dhar El Mahraz, Sidi Mohamed Ben Abdellah University, Fez, Morocco
e-mail: abdelghani.fakkahi@usmba.ac.ma

A. Ed-Dahmouny
Science and Engineering Research Laboratory (LaRSI), Faculty of Sciences and Technology, Sidi Mohamed Ben Abdellah University, Fez, Morocco

H. El-Ghazi
MPIS Group, ENSAM Laboratory, Hassan 2 University, Casablanca, Morocco

R. En-nadir (ed.), *Quantum Systems at the Nanoscale*, Lecture Notes in Nanoscale Science and Technology 45, https://doi.org/10.1007/978-3-032-06645-9_4

[16]. On the other hand, experimental studies have been done to investigate the manufacturability of nano-semiconductor materials [17–22]. The electrical and optical properties of the nano-semiconductor material are significantly altered by quantum confinement. As a result, the non-linear optical properties associated to intra-band transitions in such structures have enormous potential for use in optoelectronic and electrical devices, for this reason, several theoretical investigations on the optical nonlinearities have been done. Therefore, many authors investigate the optical properties of quantum dots (QDs) under shapes variety of the confinement potential such as Gaussian [23], Morse [24], Hulthen-Yukawa [25], Woods–Saxon [26], Pöschl–Teller [27] and Kratzer potential [28]. Kratzer potential, which was initially proposed for interactions of diatomic molecules [29] and has many applications such as chemical physics [30], molecular physics [31], information theory [32], can also be applied to quantum nanostructures [14, 33, 34]. For example, Ungan et al. [35] have investigated optical properties of *GaAs* quantum well with Kratzer-Fues potential under external factors. Spin–orbit effects of QDs in a Kratzer potential using numerical methods are studied by Nautiyal et al. [31]. Very recently, the influence of temperature and applied hydrostatic pressure on the second and third harmonic generation of spherical QD with modified Kratzer plus screened Coulomb potential are reported by Duan and co-workers [36]. Today, developments in nanostructure production technologies have made it possible to produce core/shell or multilayer nanosystems more complex than a single QD. One of the advantages of these structures is that it is possible to control the position of the carriers in the structure depending on the type of materials used in different regions. However, the boundary conditions complicate the analytical calculations of the eigenvalue and their wave functions of these structures. Therefore, numerical calculations on the physical properties of core–shell or multilayer quantum dot (MLQD) structure has piqued the interest of many authors [37–43]. The linear, nonlinear and total absorption coefficients and refractive index of a shallow donor impurity in a cylindrical MLQD under the impact of the applied hydrostatic pressure and temperature are calculated by Jaouane et al. [40]. Their obtained results show that these optical properties for $1s - 1p$, $2s - 1p$ and $1p - 1d$ transitions are sensitively dependent on the variation of QD size. So far, there are only few studies on the nonlinear optical rectification (*NOR*) of core/shell or MLQD structures in the literature. The combined effects on the NOR coefficient in $InAs/GaAs$ quantum core/shell are examined by Makhlouf et al. [44]. One of these important non-linear optical properties is the nonlinear optical rectification (*NOR*), which has received considerable attention. Moreover, various studies have been carried out on the *NOR* for different confinement potentials [10, 23, 27, 45]. One of the factors that significantly affect the physical properties of the nanoscale systems is the shape of the confining potential. The nonlinear optical rectification in different structures under the effects of the interdiffusion and electric field [9, 46, 47], the hydrostatic pressure and temperature [7], and the position-dependent mass [48] have been studied. Fakkahi et al. [10] investigated the effects of the geometrical size, external perturbations, conduction band nonparabolicity, and polaronic mass on the *NOR* of a spherical MLQD by using the finite element technique within effective mass approximation. Their findings reveal that these effects influence significantly the *NOR* profiles of considered

systems. Duque et al. [49] have investigated the effects of the applied hydrostatic pressure and electric field on the nonlinear optical rectification in a semiparabolic quantum dot. Solaimani et al. [47] have studied the effects of conduction band non-parabolicity on the nonlinear optical rectification in quantum well systems for three different structures (*GaAs*/*GaAlAs*, *GaN*/*AlN*, and *InSb*/*AlN*) by using the finite difference method. Additionally, they investigated the effect of the applied electric field on the optical rectification for the three structures mentioned above. Recently, the nonlinear optical rectification in a *GaAs* QD under the combined effect of the electron–hole pair correlation, applied hydrostatic pressure and temperature has been studied by Dahiya et al. [50]. In a more recent work, Makhlouf et al. [44] studied the combined influence of the applied electric and magnetic fields, hydrostatic pressure, temperature and the carrier's lifetime on the nonlinear optical rectification by using the finite difference method. In this chapter, we have used the finite element method to study the nonlinear optical rectification obtained in the framework of density matrix theory of a system comprised of two quantum dots of *GaAs*, one inside the other and each surrounded by layers of *GaAlAs*.. This structure is called multilayered spherical quantum dot (MSQD). Firstly, we have investigated the influence of the inner dot radius and width of the barrier on the *NOR*.. Then, we studied the effects of the external perturbations (hydrostatic pressure and temperature) on the *NOR* coefficient. In addition, the combined effects of the nonparabolicity of the con duction band and polaronic mass have been incorporated in our study. In addition, we have examined the impacts of the Kratzer potential parameters on the second harmonic generation and on the *NOR* coefficient of the considered structure for several transitions between the ground state and excited states.

2 Theory and Model

Within the effective mass approximation, the Hamiltonian of an electron bound to a hydrogenic shallow donor impurity in a *GaAs*/*GaAlAs* multilayer spherical quantum dot is expressed as follows:

$$H = -\frac{\hbar^2}{2m^*(r, P, T)}\nabla^2 + V(r, P, T) \tag{1}$$

where $m^*(r, P, T)$ is the pressure-and temperature-dependent effective mass of electron, it is defined in different regions of the MSQD as:

$$m^*(r, P, T) = \begin{cases} m_1^*(P, T) \mathit{for} GaAs \\ m_2^*(P, T) \mathit{for}\ Ga_{1-x}Al_xAs \end{cases} \tag{2}$$

The effective electron mass is related to the pressure and temperature by the following expression:

$$\frac{m_0}{m_1^*(P,T)} = 1 + E_p^\Gamma\left[\frac{2}{E_g^\Gamma(P,T)} + \frac{1}{E_g^\Gamma(P,T)+\Delta_{so}}\right] \tag{3}$$

where m_0, $E_p^\Gamma = 7.51eV$ and $\Delta_{so} = 0.341eV$ are the free electron mass, the energy related to the momentum matrix element, and the spin–orbit splitting, respectively. $E_g^\Gamma(P,T)$ is the temperature-and pressure-dependent band gap energy for *GaAs* at Γ-point determined as:

$$E_g^\Gamma(P,T) = E_g^\Gamma(0,0) + E_g^\Gamma(P) + E_g^\Gamma(T) \tag{4}$$

where $E_g^\Gamma(0,0) = 1.519eV$ is the *GaAs* energy gap in the absence of applied hydrostatic pressure and temperature, $E_g^\Gamma(P) = 0.1073$ P (eV/GPa), and $E_g^\Gamma(T) = -5.40510^{-4}T^2/(T+204)$. The effective mass of the electron in $Ga_{1-x}Al_xAs$ region is given by:

$$m_2^*(P,T) = m_1^*(P,T) + 0.083x \tag{5}$$

The limited potential $V(r)$ is defined as the some of the Kratzer-Fues potential V_K and the rectangular potential V_i. The mathematical expression of the Kratzer-Fues potential V_K is given by [49–51]:

$$V_K = -D\left(\frac{2R_0}{r} - \frac{R_0^2}{r^2}\right) \tag{6}$$

where the parameter $D = N\mathrm{V}_0$ describes the depth of the potential, and the parameter R_0 is the quantum dot radius.

The rectangular potential V_i is written as:

$$V(r,P,T) = \begin{cases} 0;\ 0 \le r \le R_1 \\ V_1;\ R_1 < r < R_2 \\ 0;\ R_2 \le r \le R_3 \\ V_2;\ R_3 < r < R_4 \end{cases} \tag{7}$$

V_i is the height of the potential barrier $(i = 1, 2)$, with $V_i = Q_C \Delta E_g^\Gamma(x,P,T)$ where Q_C is the conduction band offset which is taken to be 0.6, and $\Delta E_g^\Gamma(x,P,T)$ denotes the variation of the band gap difference between quantum well *GaAs* and the barrier matrix $Ga_{1-x}Al_xAs$ at the Γ-point, its expression in the presence of the pressure and temperature is as follows:

$$\Delta E_g^\Gamma(x,P,T) = \Delta E_g^\Gamma(x,0,0) + \Delta E_g^\Gamma(P) + \Delta E_g^\Gamma(T) \tag{8}$$

where $\Delta E_g^{\Gamma}(x, 0, 0) = 1.155x + 0.37x^2$ is the expression of the energy band-gap in the absence of hydrostatic pressure and temperature, $\Delta E_g^{\Gamma}(P) = -0.013xP$ and $\Delta E_g^{\Gamma}(T) = -0.000115xT$

The variation of the geometrical sizes of the MSQD as a function of the applied hydrostatic pressure is given by [3, 52]:

$$\mathrm{R(P) = R_0[1 - P(S_{11} + 2S_{12})]} \tag{9}$$

where R_0 is the structure size without the hydrostatic pressure, $S_{11} = 1.16 \times 10^{-2} GPa^{-1}$ and $S_{12} = 0.37 \times 10^{-2} GPa^{-1}$ are the components of the compliance tensor constants of the *GaAs*.

The *GaAs*, in fact, has a non-parabolic conduction band structure. As a result, the non-parabolicity of the conduction band must be considered in order to generate more accurate results for energy levels and donor binding energy.

It should be noted that the smallest dot radius of the QD, below which no bound state can be obtained, is given as follows [53]:

$$\mathrm{R_0} = \sqrt{\frac{\pi^2\hbar^2}{8\mathrm{m_1^*V_1}} + \frac{\hbar^2}{2\mathrm{m_2^*V_1}}\left(\frac{\mathrm{m_2^*}}{\mathrm{m_1^*}} - 1\right)^2} \tag{10}$$

Based on the k. p approximation, an energy-dependent effective mass expression is employed which takes into consideration the effects of band nonparabolicity [54]:

$$\mathrm{m_{np}^* = \left(0.067 + 0.0436E_{sub} + 0.236E_{sub}^2 - 0.147E_{sub}^3\right)m_0} \tag{11}$$

where E_{sub} is the sub-band energy expressed in meV, $\mathrm{m_0}$ is the free-electron mass. Because *GaAs* is a polar material with no center of inversion, the electron could interact with various types of phonons, via acoustic, polar optic, and piezoelectric. Thus, a polaron correction to the effective mass is required as a result of these interactions. This is known as the polaronic mass (m_{pm}^*), which is expressed as:

$$\frac{1}{\mathrm{m_{pm}^*}} = \frac{1}{\mathrm{m_{np}^*}}\left(1 - \frac{\alpha}{6}\right) \tag{12}$$

Due to the spherical symmetry, the eigenfunctions of the Hamiltonian of the system are:

$$\psi_{n,l,m}(r, \theta, \varphi) = R_{l,n}(r)Y_{l,m}(\theta, \varphi) \tag{13}$$

where $R_{l,n}(r)$ is the radial wave function and $Y_{l,m}(\theta, \varphi)$ is the spherical harmonic. n, l and m are the principal, angular momentum and azimuthal quantum numbers, respectively. From Eqs. (1) and (5), the radial part of the Schrödinger equation is:

$$\left[-\frac{m_d^*}{m_{d,b}^*}\frac{1}{r}\frac{\partial^2 r}{\partial r^2}+\frac{m_d^*}{m_{d,b}^*}\frac{l(l+1)}{r^2}-\frac{NV_0}{R_y^*}\left(\frac{2a}{r}-\frac{a^2}{r^2}\right)+\frac{V_i}{R_y^*}\right]R_{l,n}(r)=\frac{E_{nl}}{R_y^*}R_{l,n}(r) \tag{14}$$

Equation (6) is written in reduced atomic units that correspond to a unit of length of an effective Bohr radius, $a_d^* = \frac{\varepsilon_d \hbar^2}{m_d^* e^2}$ and to a unit of energy of an effective Rydberg $R_d^* = \frac{m_d^* e^4}{2\varepsilon_d^2 \hbar^2}$. Due to the existence of a Coulomb potential in the Hamiltonian Eq. (6) and due to the complexity of the system geometry, it became almost impossible to find an analytical solution. For this reason, we used the Finite Element Method (FEM), a flexible numerical method that satisfies the boundary conditions [55, 56]. It should be noted that the states 1 *s*, 2 *s*, 1*p*, and 2*p* correspond to ($n = 1$, $l = 0$), ($n = 2$, $l = 0$), ($n = 1$, $l = 1$), and ($n = 2$, $l = 1$), respectively.

2.1 Optical Rectification and Second Harmonic Generation

The following nonlinear polarization vector can describe the interaction between electromagnetic waves propagating inside a semiconductor [57, 58].

$$\mathrm{P} = \varepsilon_0 \chi_{ij}^{(1)} E_j + \varepsilon_0 \chi_{ijk}^{(2)} E_j E_k + \varepsilon_0 \chi_{ijkl}^{(3)} E_j E_k E_l + \ldots \tag{15}$$

The first term on the right-hand side of Equation represents the linear optics, $\chi^{(i)}$ is the *ith*-order nonlinear susceptibility tensor of the crystal ($i \geq 2$).

The second-order nonlinear susceptibility tensor $\chi^{(2)}$ gives rise to the phenomena of second-harmonic generation, optical rectification, Pockels electro-optic effect, etc. various nonlinear optical processes and their related dielectric susceptibility expressions (Table 1).

Second harmonic generation (SHG), optical rectification (OR), and the electro-optic Pockels effect are all nonlinear optical phenomena, but they differ in how they affect the behavior of light passing through a material.

- **Second harmonic generation (SHG):** In SHG, a material with nonlinear optical properties is used to generate a second harmonic frequency from a fundamental

Table 1 Susceptibilities and nonlinear optical processes in semiconductors

Susceptibility	*Nonlinear optical process*
$\chi_{\mathrm{ijk}}^{(2)}(-2\omega; \omega, \omega)$	Second-harmonic generation
$\chi_{\mathrm{ijk}}^{(2)}(0; \omega, -\omega)$	Optical rectification
$\chi_{\mathrm{ijk}}^{(2)}(-\omega; 0, \omega)$	Pockels electro-optic effect
$\chi_{\mathrm{ijkl}}^{(3)}(-3\omega; \omega, \omega, \omega)$	Third-harmonic generation

frequency of light. When light with a specific frequency (the fundamental) passes through the material, the nonlinear properties of the material cause the generation of a new frequency, which is exactly twice the frequency of the original light. This effect is used in many applications, including frequency doubling of lasers, and as a tool to study material properties [57].

The analytical expression of the SHG is given as:

$$\chi_{2\omega}^{(2)}(\omega) = \frac{q^3 N_e}{\varepsilon_0} \frac{M_{ij} M_{jk} M_{ki}}{\left(\hbar\omega - \Delta E_{ij} + i\hbar\Gamma_0\right)(2\hbar\omega - \Delta E_{ki} + i\hbar\Gamma_0)} \tag{16}$$

- **Optical rectification (OR):** OR is the process of generating a DC electric field from a material using an intense, pulsed laser beam. When the beam is incident on the material, the non-linear optical properties cause a tension to be generated on its surface. This tension is then rectified to produce a DC electric field. This effect is useful for generating ultrafast pulses of THz radiation, which have applications in imaging and spectroscopy.

The nonlinear optical rectification can be calculated by using the compact density-matrix approach, where the system is subjected to a spatially propagating electromagnetic wave of frequency ω, which is expressed as [31, 46]:

$$\chi_0^{(2)}(\omega) = \frac{q^3 N_e |M_{ij}|^2 \delta_{12}}{\varepsilon_0} \frac{2\Delta E^2}{\left((\Delta E - \hbar\omega)^2 + (\hbar\Gamma_0)^2\right)\left((\Delta E + \hbar\omega)^2 + (\hbar\Gamma_0)^2\right)} \tag{17}$$

where q is the absolute value of the electron charge, N_e is the MSQDs electron density, ε_0 is the dielectric constant of the vacuum, $\Delta E = E_j - E_i$ is the transition energy between the first excited state and the ground state, $M_{ij} = \Psi_j|r|\Psi_i$ is electric dipole transition matrix element and $\delta_{ij} = |M_{jj} - M_{ii}|$, Γ_0 is the relaxation rate for states $|i$ and $|j$, and $\hbar\omega$ is the incident photon energy.

- **Pockels electro-optic effect:** The electro-optic Pockels effect is a change in the refractive index of a material in response to an applied electric field. In a material with this property, an applied electric field causes a change in the refractive index, which affects the way light passes through the material. This effect is used in a variety of applications, including telecommunications to switch and modulate light signals.

In summary, the difference between these effects lies in the specific nonlinear optical behavior they exhibit. SHG generates a new frequency, OR generates a DC electric field, and the Pockels effect changes the refractive index of the material.

3 Results and Discussion

In the present study, we will discuss the numerical results of the NOR coefficient obtained by using the finite element method, for the system that is represented in Fig. 1. In all that follows, we have taken the value of the Al content $x = 0.3$ and the thickness of the first barrier to be equal to that of the second barrier. First, we have examined the effects of the geometrical sizes (δ_1 and δ_2), barrier potential height V_1, pressure P, temperature T, and the combined effect of the conduction band non-parabolicity and polaronic mass on the NOR coefficient. According to Eq. (17), the NOR resonant peak occurs at $\hbar\omega = \Delta E$, and the NOR peak magnitude is proportional to the geometrical factor $\partial_{12} M_{12}^2$. With the aim of providing a reasonable explanation of the behavior of the NOR, we have presented the effect of the structure geometrical dimensions and external perturbations on the geometrical factor $\partial_{12} M_{12}^2$ and the energy of the initial and final states E_1 and E_2, respectively.

3.1 The Effect of the Geometric Sizes

In Fig. 2, we present the variation of the nonlinear optical rectification with the incident photon energy $\hbar\omega$ for various values of the inner dot radius $\delta_1 = 0.4, 0.6$ and $1\ a_d^*$, and for fixed values of the barrier width $\delta_2 = 0.4 a_d^*$ and outer QD radius

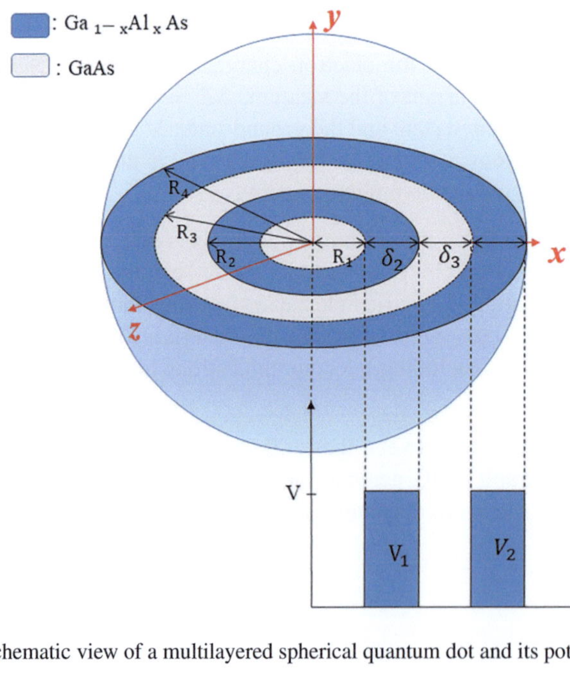

Fig. 1 Schematic view of a multilayered spherical quantum dot and its potential profile

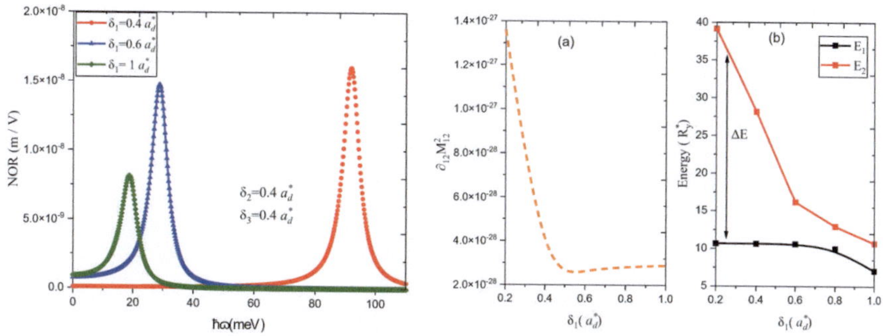

Fig. 2 The nonlinear optical rectification as a function of the incident photon energy for different values of the inner dot radius (left panel), and the variation of the geometrical factor **a** and the energy levels of the ground state E_1 and first excited state E_2 **b** as a function of the inner dot radius δ_1(right panel)

$\delta_3 = 0.6a_d^*$. As can be seen from the figure, for a fixed value of the inner dot radius, the NOR increases, reaches a maximum which occurs at $\hbar\omega = \Delta E$ and then decreases as the photon energy increases. This behavior is analogous to that found in the literature [26, 45, 46]. Moreover, it is clearly seen that the NOR is strongly dependent on the inner QD radius δ_1, i.e., when the latter increases the peak position of the NOR displaces to lower energy and the NOR peak amplitude reduces. This can be explained by the decrement in the energy difference between the first excited and the ground energy states observed in the Fig. 2b, due to the diminution of the confinement of the carriers which leads to a redshift of the resonant peak. On the other hand, the decline in the NOR's resonant peak intensity is due to the $\partial_{12}.M_{12}^2$ decreasing as the inner dot radius increases (see Fig. 2a). These findings match well with previous studies using different technical in QDs with diverse shapes [35, 46].

Figure 3 reveals the graph of the NOR as a function of the incident photon energy $\hbar\omega$ for several values of the barrier width $\delta_2(0.2a_d^*, 0.4a_d^*$, and $0.6a_d^*)$, the outer QD radius is taken to change as:$\delta_3 = 1 - \delta_2$, and the radius of the inner dot is fixed at $\delta_1 = 0.4a_d^*$. One may see from Fig. 3 that as the barrier thickness grows, the NOR's peak position is displaced to the lower energies (redshift). The main physical reason for this phenomenon is that the coupling effect between the dots becomes progressively weaker as the barrier width becomes gradually larger [40]. As a result, the difference in energy between the electronic states E_1 and E_2 becomes narrower (as seen in Fig. 3b), which leads to the redshift of the resonant peak. Furthermore, the intensity of the resonant peak of the NOR decreases with increasing barrier width. This behavior can be attributed to the sharp decrease in the product of matrix elements $\partial_{12}.M_{12}^2$ with increasing barrier thickness observed in Fig. 3a. It can be concluded that the amplitude and position of the resonant peak depend on the barrier thickness.

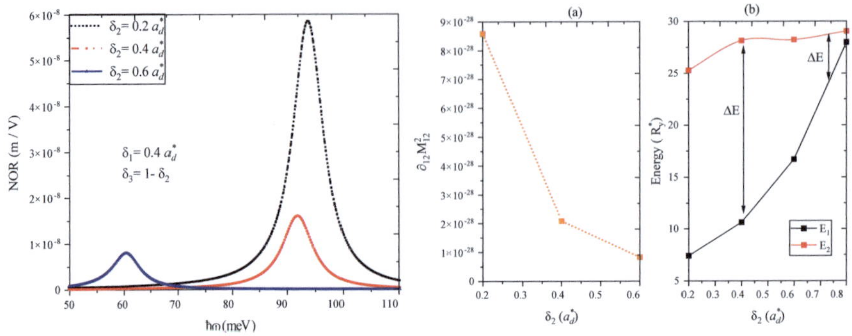

Fig. 3 The nonlinear optical rectification as a function of the incident photon energy for different values of the barrier width (left panel), and the variation of the geometrical factor **a** and the energy levels of the ground state E_1 and first excited state E_2 **b** as a function of the barrier thickness δ_2(right panel)

4 The Effect of the Confinement Potential

Figure 4 depicts the graph of the NOR as a function of incident photon energy, for different values of the barrier potential height V_1 ($V_1 = V_2, V_1 = 1.25V_2, V_1 = 1.5V_2$ and $V_1 = 2V_2$). While the barrier thickness, inner QD radius, and outer QD radius are fixed at $\delta_2 = 0.6a_d^*$ and $\delta_1 = \delta_3 = 0.4a_d^*$, respectively. The most striking characteristic of the graph is how the magnitude of the NOR peak decreases as the confinement potential height increases V_1. The physical reason for this behavior is that the matrix element product $\partial_{12}.M_{12}^2$ decreases significantly as the potential confinement height V_1 increases (see Fig. 4a). It is also observed from the figure that the resonant peak position shifts toward higher energies (blue shift) as the barrier potential becomes larger because the quantum confinement of the carriers increases with the increment of the potential height. This is confirmed by the enhancement of the sub-bands energies E_1 and E_2 when the barrier potential height increases, as noted in Fig. 4b. These findings indicate that the confinement potential should be small in order to achieve a significant NOR coefficient. These results are in good agreement with those of Refs. [20, 23, 25].

4.1 The Effects of the Hydrostatic Pressure and Temperature

In order to study the influence of the hydrostatic pressure on the nonlinear optical rectification, we have plotted in Fig. 5, the variation of the NOR coefficient as a function of the incident photon energy $\hbar\omega$, for different values of applied hydrostatic pressure $P = 0, 2, 3$ and $4GPa$, while the temperature is kept at $T = 300K$, and for a fixed set of parameters $\delta_2 = 0.6a_d^*$ and $\delta_1 = \delta_3 = 0.4a_d^*$. As it can be revealed from the figure, the hydrostatic pressure has a considerable effect on the NOR coefficient.

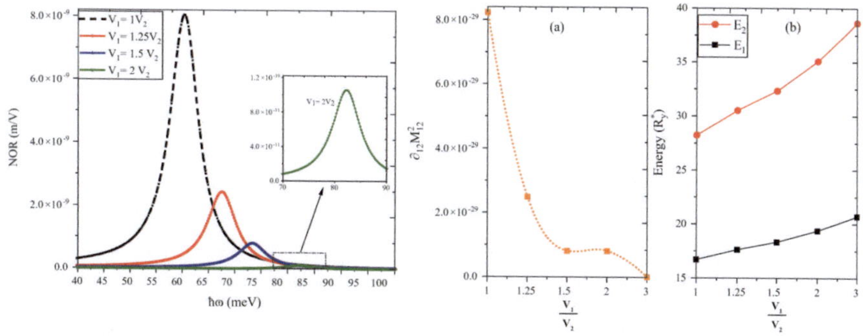

Fig. 4 The nonlinear optical rectification as a function of the incident photon energy for different values of potential height V_1(left panel), and the variation of the geometrical factor **a** and the energy levels of the ground state E_1 and first excited state E_2 **b** as a function of the potential height V_1 (right panel)

As we know from the literature, the inclusion of the hydrostatic pressure yields a decreasing (increasing) in the dielectric constant and the size of the dot (the electron effective mass), which leads to a decrement in the energy levels of the both electronic states (initial and final state), as illustrated in Fig. 5b. As a result, the application of hydrostatic pressure causes a red-shift of the resonant peak. Furthermore, the amplitude of the resonant peak diminishes as the applied hydrostatic pressure increases. These results can be interpreted physically by the sharp decrease observed in the geometrical factor $\partial_{12}.M_{12}^2$ with hydrostatic pressure, as illustrated in Fig. 5a. These results are in good agreement with previous work in different structures [21, 46, 48].

Figure 6 displays the NOR coefficient as a function of $\hbar\omega$ for different temperature values, $T = 300, 400$ and $500K$, while the hydrostatic pressure P is kept at $P =$

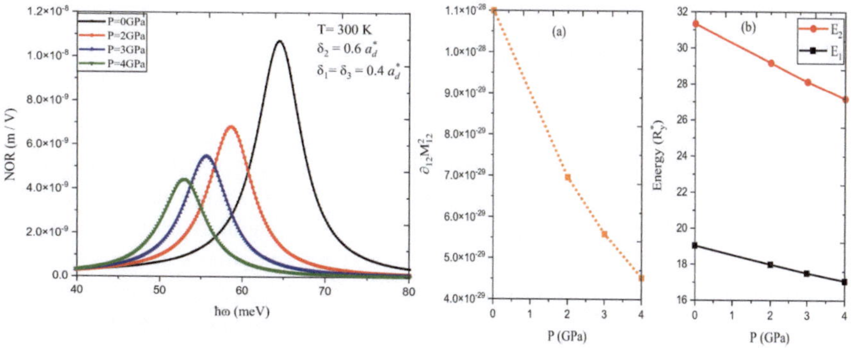

Fig. 5 The nonlinear optical rectification as a function of the incident photon energy for different values of hydrostatic pressure (left panel), and the variation of the geometrical factor **a** and the energy levels of the ground state E_1 and first excited state E_2 **b** as a function of the applied hydrostatic pressure (right panel)

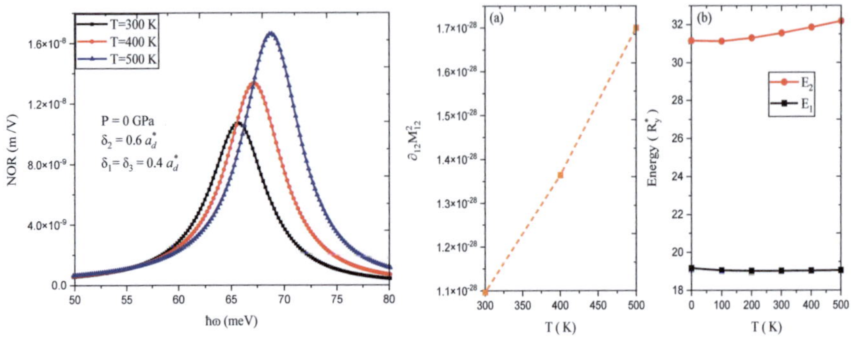

Fig. 6 The nonlinear optical rectification as a function of the incident photon energy for different values of temperature, and the variation of the geometrical factor **a** and the energy levels of the ground state E_1 and first excited state E_2 **b** as a function of the temperature

$0GPa$, with the following fixed parameters, $\delta_1 = \delta_3 = 0.4a_d^*$ and $\delta_2 = 0.6a_d^*$. It is clearly seen from the figure that the magnitude and position of the resonant peak are obviously dependent on the temperature. In contrast to the hydrostatic pressure effect, increasing temperature values cause the resonant peak position to shift to higher energies and the peak magnitude to rise. The physical reasons are that the energy separation ΔE increases with the increase of the temperature value, as seen in Fig. 6b, which causes a "blue shift" of the NOR resonance peak position. Moreover, it can be clearly seen that the NOR peak amplitude increases with increasing temperature. This enhancement of the NOR peak magnitude can be explained by the rapid increase of the geometrical factor $\partial_{12}.M_{12}^2$ with increasing temperature, as illustrated in Fig. 6a. The general behavior of the NOR coefficient under the effect of the temperature is in agreement with that observed in Refs. [34, 46].

4.2 *The Effects of Nonparabolicity of the Band and Polaronic Mass*

In Fig. 7 the NOR coefficient is plotted against the incident photon energy $\hbar\omega$, for a fixed parameter $\delta_1 = \delta_3 = 0.4a_d^*$ and $\delta_2 = 0.6a_d^*$ by taking into account the three following effects: parabolic band effect, band nonparabolicity effect, and the combined effects of the band nonparabolicity and polaronic mass. It is clearly seen from the figure that the introduction of the conduction band nonparabolicity and polaronic mass effects lead to a reduction in the peak magnitude of the NOR. This is due to the fact that both effects diminish the geometrical factor $\partial_{12}.M_{12}^2$. The figure also indicates that when the the nonparabolic conduction band and mass polaronic effect are taken into account, the resonant peak location shifts towards lower energies (red-shift). This is because the conduction band nonparabolicity and polronic mass

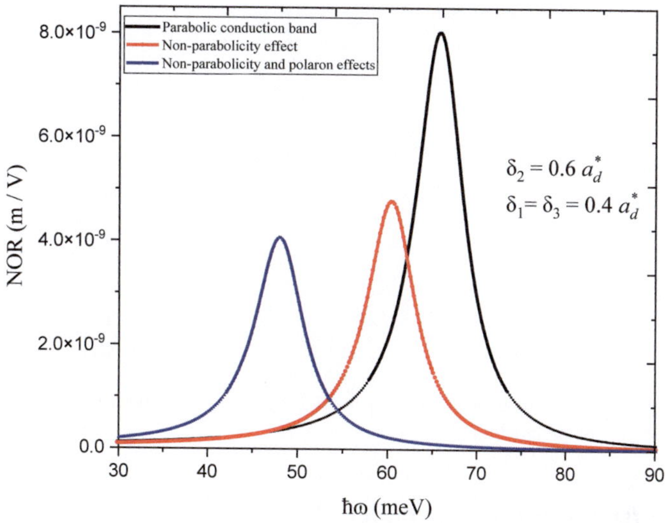

Fig. 7 The optical rectification coefficient as a function of the incident photon energy taking into account the effect of conduction band non-parabolic and polaronic mass

effects reduce the energy separation ΔE. Furthermore, when the effects of conduction band nonparabolicity and polaronic mass are simultaneously introduced in the *GaAs*/*GaAlAs* MSQD, competition effects appear. The figure reveals that the polaronic mass has little influence on the resonant peak intensity for the used MSQD heterostructure geometrical sizes. It has been shown that the polaronic mass effect dominates over the nonparanolic conduction band one as the system becomes thick (weak quantum confinement) [4].

4.3 The Excited States

In this section, we present the nonlinear optical rectification coefficient (NOR) and second harmonic generation (SHG) coefficients in a multilayered spherical quantum dot under Kratzer potential confinement, which is presented in Fig. 8. We used atomic units throughout the numerical calculations. The fixed parameters that we utilized for our numerical calculations are as follows: $\hbar\Gamma_0 = 0.4\text{meV}$, $\text{x} = 0.3, \delta_1 = \text{R}_0, \delta_2 = 0.6a_d^*, \delta_3 = 0.4a_d^*$, $\text{R}_\text{y}^* = 5.24\text{meV}$ and $a_d^* = 10.4\,\text{nm}$.

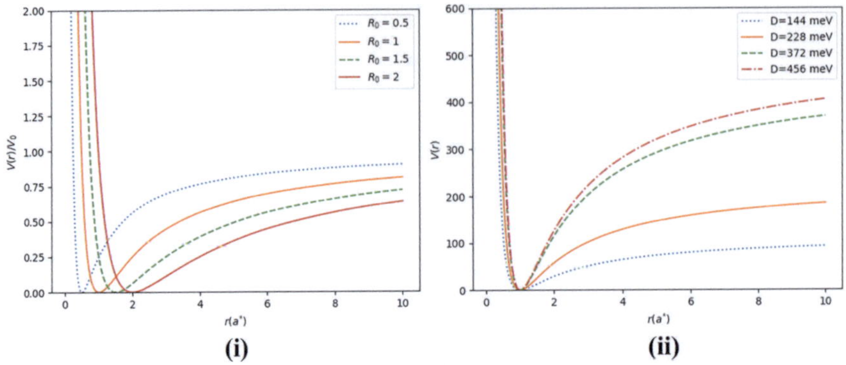

Fig. 8 The profile of the Kratzer potential **a** for different values of the parameters R_0 and **b** for different values of the parameters D

4.4 The Effect of the Parameter $\mathbf{R}_0$

Figure 11a displays the behavior of the NOR coefficient as a function of the incident photon energy $\hbar\omega$ for several values of the parameter R_0 and for the $1s - 1p$ transition. As is well known, the NOR coefficient has a resonance energy value, which corresponds to a maximum value for the energy of the incident photon equal to the energy difference between quantum states. It is obvious that the change of R_0 has great influence on the intensities of the resonant peaks. As can be seen from Fig. 11a, the NOR coefficient peaks experience a red-shift with increasing R_0 parameter. The physical origin is that, the separation between the $1s$ and $1p$ energy levels becomes narrower with increasing R_0 (see Fig. 10a). We can see that the increase of R_0 parameter enhances the maximum value of the NOR coefficient. This situation can be understood with the geometrical factor $\partial_{ij} M_{ij}^2$ given in Fig. 10b for $1s - 1p$ transition. To understand the behavior of NOR depending on the $R0$ parameter, we should be discussed the dependence of this parameter on the transition energy and matrix element between different quantum states. The variation of transition energy of the system between different energy levels versus the parameter R_0 is shown in Fig. 10a. This figure clearly shows that the transition energy for $1p - 2s$ decreases monotonously and rapidly with increasing R_0 than that of $1s - 1p$ transition. In contrast, the behavior of $2s - 2p$ transition is non-monotonic, i.e., it increases as R_0 becomes larger, and then it reaches a maximum value and diminishes with further increases of R_0. In all transitions, the difference in energy levels are smaller for large R_0 values. Also, it is clear that $R0$ parameter has more influence on transition between higher energy levels. The dipole matrix element can give important information to understand the variation of maximum in NOR coefficients. Figure 10b shows the geometrical factor, $\partial_{ij} M_{ij}^2$ against R_0 for different transitions between energy levels for constant $N = 1 (D = V_0)$ and other parameters. It is seen that the dipole matrix elements for $1s - 1p$ and $1p - 2s$ transitions increase with increasing R_0. While this increment is almost linear for $1s - 1p$ transition, it occurs rapidly after $R_0 = 0.4a^*$, value for $1p - 2s$ transition. However, the behavior of the dipole transition matrix element for $2s - 2p$ is quite different from other transitions. In this transition, matrix element decreases with R_0 and exhibits minimization at $R_0 = 0.6a^*$, then it displays an enhancement with increase in R_0. In Fig. 11b, we have plotted the NOR coefficient of $1p - 2s$ transition as a function of the incident photon energy for several values of R_0. It is clearly seen that the behavior of the NOR coefficient is similar to that reported in Fig. 11a. But, when these two figures are compared, it is seen that the curve of NOR coefficient for $1p - 2s$ is sharper than that of $1s - 1p$ transition, and the effect of the parameter R_0 is less pronounced for the last transition.

Figure 12 shows the NOR coefficient dependence on the incident photon energy for $2s - 2p$ transition and for different values of R_0 With the increase of R_0, the maximum value of NOR coefficient decreases to a minimum value and then increases due to non-monotonic behavior of the geometrical factor as discussed in Fig. 10b (filled circle red curve). We find that, when $R_0 = 0.6a^*$, the smallest peak value of NOR coefficient has a minimum value in Fig. 12. In addition, from this figure, we have also observed that the peak value of NOR coefficient move to the right side (blue shift) up to $R_0 = 0.4a^*$ and it moves to the left side (red shift) for $R_0 > 0.4a^*$. This behavior can be explained by the variations of $2s - 2p$ transition energy versus the paramter R_0 as shown in Fig. 10a. Furthermore, when we compared the amplitude of the peaks corresponding to $R_0 = 0.4a^*$ and $R_0 = 0.8a^*$ we observed that the magnitude of the peak increased owing to wave functions overlapping, as shown in Figs. 9a and b. As it can be seen from Fig. 12, the NOR coefficient is affected drastically by R_0 parameter and this effect is strongly dependent on the R_0 parameter selected.

4.5 *The Effect of the Parameter D*

For four different values of N ($N = 1, 2, 3$ and 4), in Fig. 14b, we show the variation of NOR coefficient for $2s - 2p$ transition as a function of the incident photon energy. According to this figure, the peak position shifts to higher photon energy values up to a critical N value ($N = 2$), and then it exhibits a red-shift. This behavior is explicable by considering the variation of the transition energy with N for $2s - 2p$ shown in Fig. 13b. Also, the magnitude of the NOR coefficient increases with increasing N, which is consistent with the behavior of the geometrical factor of $2s - 2p$ show in Fig. 14a. In Fig. 15b, in order to investigate the influence of the parameter N on the NOR coefficient, we draw the later for $1p - 2s$ transition as a function of the excitation photon energy for four different values of the parameter N ($N = 1, 2, 3$ and 4). As seen in this figure, the NOR coefficient exhibits blue shift as N becomes higher. This displacement of the peak position is due to the increment of the transition energy with increasing the parameter N (Fig. 13a). In addition, the resonant peak intensity of the NOR coefficient first decreases and then increases with the parameter N. The main reason for this change is due to the behavior of the geometrical factor for $1p - 2s$ transition with the depth of the potential shown in Fig. 14a, as well as

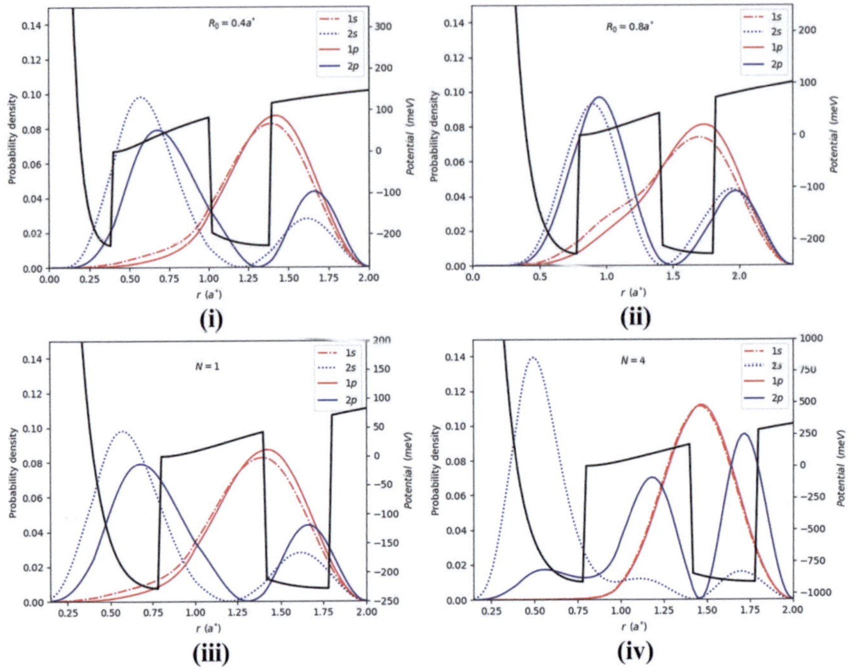

Fig. 9 The curves **a** and **b** are the probability density of the electron for $R0 = 0.4a$ and $R0 = 0.8a$ respectively. **c** and **d** for N = 1 and N = 4, respectively

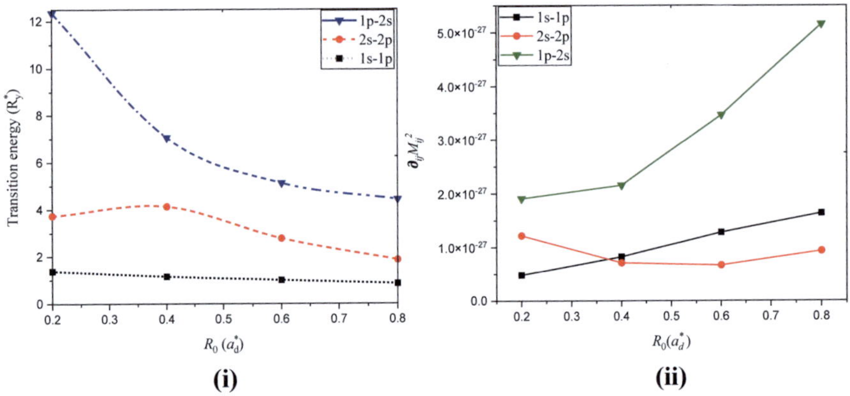

Fig. 10 **a** The transition energy and **b** geometrical factor as a function of the core radius for different transition $1s - 1p$, $1p - 2s$, and $2s - 2p$

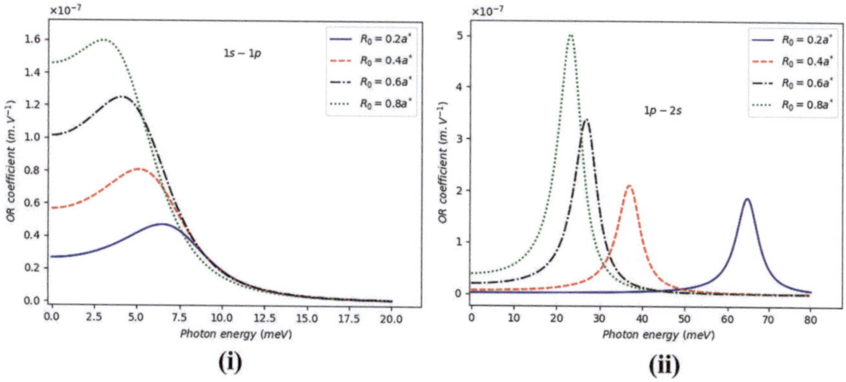

Fig. 11 The nonlinear optical rectification as a function of the photon energy for the transition **a** 1s-1p and **b** 1p-2s

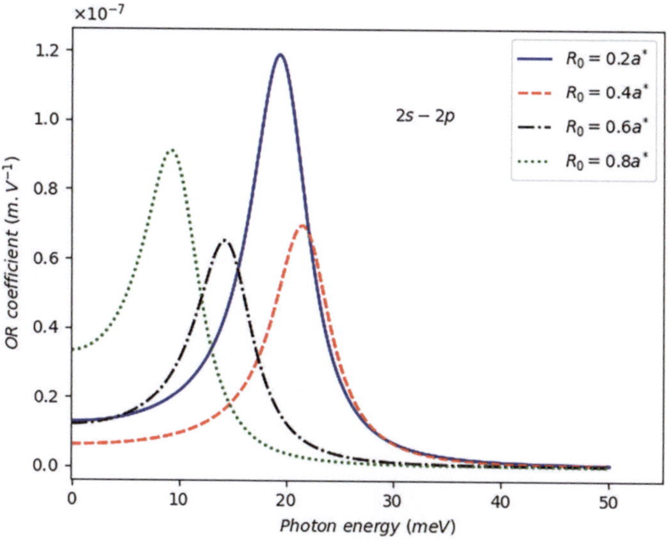

Fig. 12 The nonlinear optical rectification as a function of the incident photon energy for the transition $2s - 2p$ and for different values of the parameter R_0

the overallaping of the both wave functions decrease with N as seen in Figs. 9c and d in the well region.

Figure 15a demonstrates the dependence of the NOR coefficient on the energy of the incident photon $\hbar\omega$ for four values of the parameter N with $1s - 1p$ transition. The N parameter plays a reducing effect on the magnitude of NOR coefficient for $1s - 1p$ transition. The larger values of NOR coefficient are obtained for smaller values of the parameter N, i.e. in the case that $N = 1$. Furthermore, we observe that the decrease in the peak value is quite significant from $N = 1$ to $N = 2$, whereas, for higher N values,

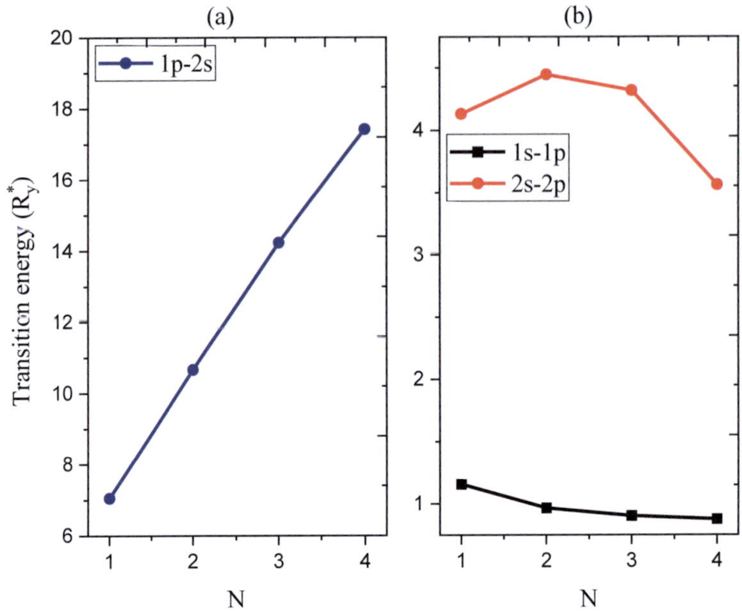

Fig. 13 The transition energies as a function of the parameter N for three allowed transitions $1s-1p$, $1p-2s$ and $2s-2p$

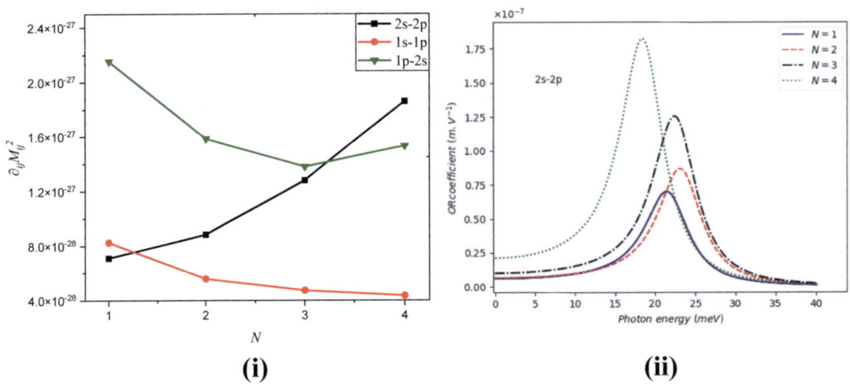

Fig. 14 **a** The geometrical factor $\partial_{ij}M_{ij}^2$ of several transition as a function of the parameter N and **b** the nonlinear optical rectification as a function of the incident photon energy for the transition $2s$–$2p$ and for different values of the parameter N

the decrease in peak values is less and the curves approach each other. In Fig. 14a, we report the geometrical factor of several transitions as a function of the parameter N with constant $R_0 = 0.4a^*$ value. This figure clearly shows that the geometrical factor of the $1s-1p$ transition decreases with N parameter. In contrast, the depth of the

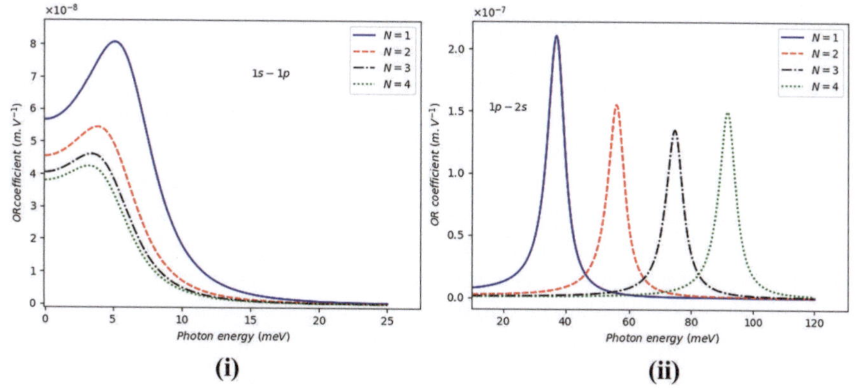

Fig. 15 The nonlinear optical rectification as a function of the incident photon energy for the transition **a** $1s - 1p$ and **b** $1p - 2s$ for different values of the parameter N

potential influences the distributions of wavefunctions, which affects the magnitude of the $2s - 2p$ matrix element at last. When comparing the effect of N parameter on the matrix element, the $1p-2s$ matrix element has a different behavior than two other transitions. The matrix element of $1p-2s$ decreases with the increase of N parameter, dropping to a minimum value and increases when N exceeds the value $N=3$. We can also say that the resonant peaks of NOR coefficients shift to the lower energy region (red shift) with increasing the parameter N. To understand the physical reason of this shift, we have presented the transition energies as a function of the parameter N in Fig. 13a (left panel) and b (right panel) are plotted for $1p-2s$ transition and $1s-1p$, and $2s - 2p$ transitions, respectively. When we look at Fig. 13, the characteristic behaviors of the transition energies for different quantum levels with increasing N are not similar. For $1p-2s$ transition, the transition energy increases linearly with depth of the potential. On the other hand, with the increasing of N parameter, transition energy of between $1p$ and $1s$ decreases monotonously. However, the change in N parameter causes the $2s - 2p$ transition energy to behave differently to the other transitions.

4.6 The Second Harmonic Generation

Figure 17a presents the second harmonic generation (SHG) curve versus the incident photon energy for different R_0 values. The SHG coefficient possesses two resonant peaks, the major one and the minor one. These peaks are related to transition energy and they occur at the $\hbar\omega = E_{01}$ and $\hbar\omega = \frac{E_{20}}{2}$. As the figure ravels, when R_0 increases, the value of the SHG coefficient increases obviously. In order to understand the relation between the resonant peak of SHG and R_0 parameter, the product of the matrix elements product $M_{01}M_{12}M_{20}$ are drawn in Fig. 16a. For small R_0 values, the dipole matrix element increases slightly, while for values larger than $R_0 = 0.4a^*$ it increases rapidly. The variation in the peak values of the SHG is pronounced for higher values of the parameter R_0 and these results are in agreement with Fig. 16a. Furthermore, we can also see from the Fig. 17a that by increasing R_0 parameter, the major resonant peak shift to lower energies and the both peaks get closer to each other. The SHG coefficients for different values of N parameter are presented in Fig. 17b versus the incident photon energy. As seen in Fig. 17b, the peak intensity of the SHG coefficient diminishes when the parameter N increases, i.e. the potential becomes more deeper, the reason for this is that $M_{01}M_{12}M_{20}$ decreases when N increases (see Fig. 16b). Moreover, an increment in N causes a blue shift of the resonant peak of SHG coefficient and the peaks get more separated with N values.

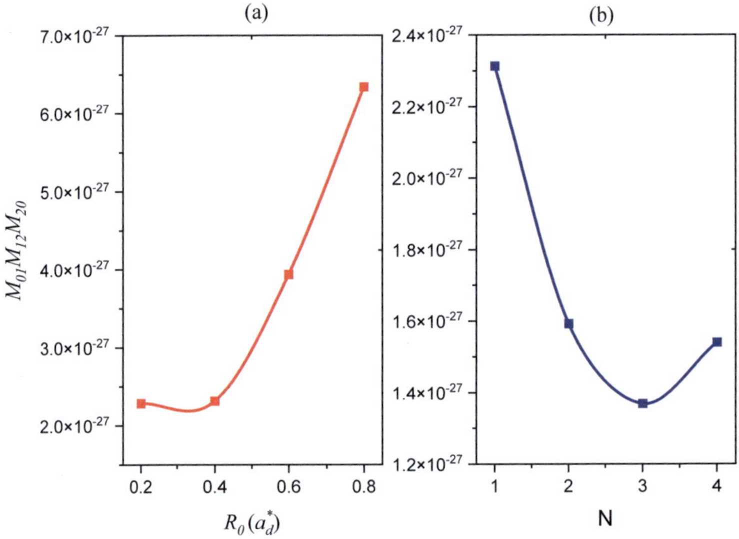

Fig. 16 The product of the element matrix $M_{01}M_{12}M_{20}$ as a function of the parameter R_0 **a** and the parameter N **b**

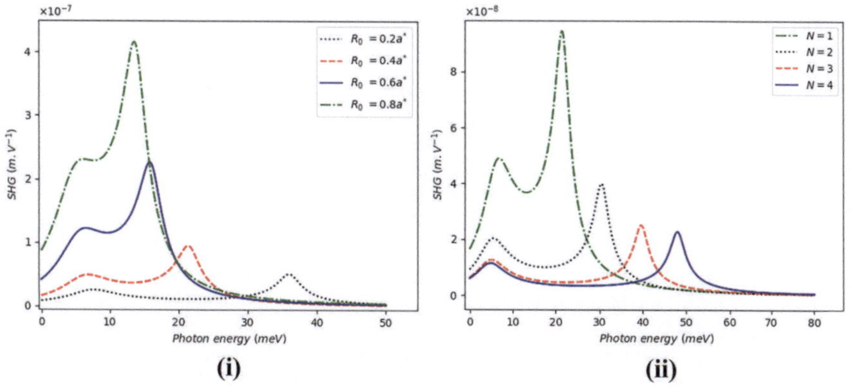

Fig. 17 The second harmonic generation as a function of the incident photon energy **a** for different values of the parameter R_0 and **b** for different values of the parameter N

5 Conclusion

In this chapter, we investigated the effects of the variation of the structure sizes (inner dot radius and barrier width), hydrostatic pressure and temperature on the nonlinear optical rectification in a $GaAs/GaAlAs$ MSQD. Furthermore, we have investigated the nonlinear optical rectification coefficient (NOR) for various transitions between different quantum levels and second-harmonic generation (SHG) coefficients in $GaAs/AlGaAs$ spherical MLQD under Kratzer potential confinement. To this end, the eigenvalues of the ground-state and excited states and their corresponding eigenfunctions are calculated within effective mass approximation and by using finite element technique. The obtained numerical results reveal that the resonance peak of the OR coefficient depends strongly on the geometrical size and the external perturbation (hydrostatic pressure, temperature). In addition, our findings show that the combined effect of the conduction band non-parabolicity and the polaronic mass affect dramatically the NOR of the MSQD. In addition, the numerical results obtained show that both depth of the potential (N) and the minimum point of potential (R_0) parameters affect drastically these optical properties. Generally, we have observed that the maximum value of NOR coefficients increases (decreases) as R_0 (N) parameter increases. Moreover, the peaks of NOR coefficient can undergo a blue or red shift depending on these parameters in transitions between different quantum levels. Also, it is found that the NOR for $2s-2p$ presents a different behavior as a function of the photon energy than that of other transitions. We hope that this present investigation stimulates future experimental and theoretical studies related to nonlinear optical properties of nanostructures.

References

1. K. Imamura, Y. Sugiyama, Y. Nakata, S. Muto, N.Y.N. Yokoyama, New optical memory structure using self-assembled inas quantum dots. Jpn. J. Appl. Phys. **34**(11A), L1445 (1995)
2. F.D.P. Alves, G. Karunasiri, N. Hanson, M. Byloos, H. Liu, A. Bezinger, M. Buchanan, Nir, mwir and lwir quantum well infrared photodetector using interband and intersubband transitions. Infrared Phys. Technol. **50**, 182–186 (2007)
3. M. Bass, P. Franken, J. Ward, G. Weinreich, Optical rectification. Phys. Rev. Lett. **9**(11), 446 (1962)
4. C. Duque, M. Mora-Ramos, C. Duque, Hydrostatic pressure and electric field effects and nonlinear optical rectification of confined excitons in spherical quantum dots. Superlattices Microstruct. **49**(3), 264–268 (2011)
5. G. Liu, K. Guo, J.-H. Wu, Polaron effects on the optical rectification in cylindrical quantum dots with electric and magnetic fields. Superlattices Microstruct. **57**, 102–114 (2013)
6. B. Vaseghi, G. Rezaei, V. Azizi, S. Azami, Spin–orbit interaction effects on the optical rectification of a cubic quantum dot. Phys. E **44**(7–8), 1241–1243 (2012)
7. L. Bouzaïene, R.B. Mahrsia, M. Baira, L. Sfaxi, H. Maaref, Hydrostatic pressure and temperature effects on nonlinear optical rectification in a lens shape inas/gaas quantum dot. J. Lumin. **135**, 271–275 (2013)
8. H. Bahramiyan, Electric field and impurity effect on nonlinear optical rectification of a double cone like quantum dot. Opt. Mater. **75**, 187–195 (2018)
9. M. Solaimani, H. Moghadam, Effects of interdiffusion and electric field on the optical rectification coefficient of $GaAs/Al_wGa_{1-w}As$ systems: crossover from single to multiple quantum wells. Appl. Phys. A **126**, 278 (2020)
10. A. Fakkahi, A. Sali, M. Jaouane, R. Arraoui, A. Ed-Dahmouny, Investigation of the nonlinear optical rectification coefficient in a multilayered spherical quantum dot. Opt. Mater. **132**, 112752 (2022)
11. A. Fakkahi, H. Dakhlaoui, A. Sali, M. Jaouane, R. Arraoui, K. El-bakkari, A. Ed-Dahmouny, Pressure, temperature and electric field effects on the photoionization cross section in a multilayered spherical quantum dot. Eur. Phys. J. Plus **137**(11), 1244 (2022)
12. A. Fakkahi, M. Kirak, M. Jaouane, A. Sali, A. Ed-Dahmouny, K. El-Bakkari, R. Arraoui, The nonlinear optical rectification and second harmonic generation of a single electron confined in a multilayer spherical quantum dot. Opt. Quantum Electron. **55**(5), 476 (2023)
13. M. Jaouane, A. Sali, A. Fakkahi, R. Arraoui, A. Ed-Dahmouny, F. Ungan, Photoionization cross section of donor single dopant in multilayer quantum dots under pressure and temperature effects. Phys. E: Low-Dimensional Syst. Nanostruct. **144**, 115450 (2022)
14. E. Hakobyan, Nonlinear optical properties of cylindrical quantum dot with kratzer confining potential in the presence of axial homogeneous electric field, in *Journal of Physics: Conference Series*, vol. 1326, (IOP Publishing, 2019) p. 012008
15. C. Chang, X. Li, Y. Ma, Nonlinear optical rectification of gaas/alηga1ηas quantum dots system with hulthén-yukawa potential. Physica B **639**, 413975 (2022)
16. A. Portacio, L. Cano, D. Rasero, Optical rectification in self-assembled quantum dots: The role of incoherent pumping. Superlattices Microstruct. **156**, 106937 (2021)
17. K. Watanabe, N. Koguchi, Y. Gotoh, Fabrication of GaAs quantum dots by modified droplet epitaxy. Jpn. J. Appl. Phys. **39**, L79–L81 (2000)
18. S. Sanguinetti, K. Watanabe, T. Tateno, M. Gurioli, P. Werner, M. Wakaki, N. Koguchi, Modified droplet epitaxy GaAs/AlGaAs quantum dots grown on a variable thickness wetting layer. J. Cryst. Growth **253**(1–4), 71–76 (2003)
19. D.R. Matthews, H.D. Summers, P.M. Smowton, M. Hopkinson, Experimental investigation of the effect of wetting-layer states on the gain–current characteristic of quantum-dot lasers. Appl. Phys. Lett. **81**, 4904–4906 (2002)
20. Z.M. Wang, K. Holmes, Y.I. Mazur, G.J. Salamo, Fabrication of (In, Ga)As quantum-dot chains on GaAs(100). Appl. Phys. Lett. **84**, 1931–1933 (2004)

21. M. Usman, S. Heck, E. Clarke, P. Spencer, H. Ryu, R. Murray, G. Klimeck, Experimental and theoretical study of polarization-dependent optical transitions in InAs quantum dots at telecommunication-wavelengths (1300–1500 nm). J. Appl. Phys. **109**, 104510 (2011)
22. Y.-C. Tsai, M.-Y. Lee, Y. Li, M.M. Rahman, S. Samukawa, Simulation study of multilayer Si/SiC quantum dot superlattice for solar cell applications. IEEE Electron Device Lett. **37**(6), 758–761 (2016)
23. S. Pal, M. Ghosh, Tailoring nonlinear optical rectification coefficient of impurity doped quantum dots by invoking Gaussian white noise. Opt. Quantum Electron. **48**, 1–18 (2016)
24. F. Ungan, M. Bahar, M. Barseghyan, L. Perez, D. Laroze, Effect of intense laser and electric fields on nonlinear optical properties of cylindrical quantum dot with morse potential. Optik **236**, 166621 (2021)
25. X. Li, C. Chang, Nonlinear optical properties of gaas/alηga1-ηas quantum dots system with hulthén-yukawa potential. Opt. Mater. **131**, 112605 (2022)
26. O. Aytekin, S. Turgut, V.Ü. Ünal, E. Akşahin, M. Tomak, Nonlinear optical properties of a Woods–Saxon quantum dot under an electric field. Phys. E **54**, 257–261 (2013)
27. O. Aytekin, S. Turgut, M. Tomak, Nonlinear optical properties of a pöschl–teller quantum well under electric and magnetic fields. Phys. E **44**(7–8), 1612–1616 (2012)
28. K. Batra, V. Prasad, Spherical quantum dot in kratzer confining potential: study of linear and nonlinear optical absorption coefficients and refractive index changes. Eur. Phys. J. B. **91**, 1–11 (2018)
29. A. Kratzer, Die ultraroten rotationsspektren der halogenwasserstoffe. Z. Phys. **3**, 289–307 (1920)
30. G. Van Hooydonk, Gauge symmetry, chirality and parity effects in four-particle systems: coulomb's law as a universal function for diatomic molecules. Spectrochim. Acta Part A Mol. Biomol. Spectrosc. **56**(12), 2273–2331 (2000)
31. D.A. Nugraha, A. Suparmi, C. Cari, B.N. Pratiwi, Asymptotic iteration method for solution of the kratzer potential in d-dimensional Klein-Gordon equation, in *Journal of Physics: Conference Series*, vol. 820, (IOP Publishing, 2017), p. 012014
32. S. Najafizade, H. Hassanabadi, S. Zarrinkamar, Nonrelativistic Shannon information entropy for Kratzer potential. Chin. Phys. B **25**(4), 040301 (2016)
33. R. Khordad, Confinement of an exciton in a quantum dot: effect of modified kratzer potential. Indian J. Phys. **87**, 623–628 (2013)
34. A. Dehyar, G. Rezaei, A. Zamani, Electronic structure of a spherical quantum dot: effects of the Kratzer potential, hydrogenic impurity, external electric and magnetic fields. Phys. E **84**, 175–181 (2016)
35. F. Ungan, J.C. Martínez-Orozco, R. Restrepo, M.E. Mora-Ramos, The nonlinear optical properties of gaas-based quantum wells with kratzer–fues confining potential: role of applied static fields and non-resonant laser radiation. Optik **185**, 881–887 (2019)
36. Y. Duan, X. Li, C. Chang, Z. Zhao, The second and third-harmonic generation of spherical quantum dots under modified kratzer plus screened coulomb potential. Physica B **639**, 413941 (2022)
37. F.K. Boz, S. Aktas, A. Bilekkaya, S.E. Okan, Geometric effects on energy states of a hydrogenic impurity in multilayered spherical quantum dot. Appl. Surf. Sci. **255**(13–14), 6561–6564 (2009)
38. M. Karimi, G. Rezaei, M. Nazari, Linear and nonlinear optical properties of multilayered spherical quantum dots: effects of geometrical size, hydrogenic impurity, hydrostatic pressure and temperature. J. Lumin. **145**, 55–60 (2014)
39. M. Tanhaei, G. Rezaei, Hydrogenic impurity, external electric and magnetic fields effects on the nonlinear optical properties of a multi-layer spherical quantum dot. Superlattices Microstruct. **98**, 29–36 (2016)
40. M. Jaouane, A. Sali, A. Ezzarfi, A. Fakkahi, R. Arraoui, Study of hydrostatic pressure, electric and magnetic fields effects on the donor binding energy in multilayer cylindrical quantum dots. Phys. E **127**, 114543 (2021)
41. A. Fakkahi, M. Kirak, A. Sali, Effect of impurity position and electric field on the optical absorption coefficients and oscillator strength in spherical multilayer quantum dot. Eur. Phys. J. Plus **137**(9), 1–19 (2022)

42. V. Holovatsky, M. Chubrey, O. Voitsekhivska, Effect of electric field on photoionisation cross-section of impurity in multilayered quantum dot. Superlattices Microstruct. **145**, 106642 (2020)
43. M. Chubrei, V. Holovatsky, C. Duque, Effect of magnetic field on donor impurity-related photoionisation cross-section in multilayered quantum dot. Phil. Mag. **101**(24), 2614–2633 (2021)
44. D. Makhlouf, M. Choubani, F. Saidi, H. Maaref, Applied electric and magnetic fields effects on the nonlinear optical rectification and the carrier's transition life-time in inas/gaas core/shell quantum dot. Mater. Chem. Phys. **267**, 124660 (2021)
45. İ Karabulut, H. Şafak, Nonlinear optical rectification in semiparabolic quantum wells with an applied electric field. Phys. B: Condensed Matter. **368**(1–4), 82–87 (2005)
46. İ Karabulut, H. Şafak, M. Tomak, Nonlinear optical rectification in asymmetrical semiparabolic quantum wells. Solid State Commun. **135**, 735–738 (2005)
47. M. Solaimani, S.M. Ali Aleomraninejad, L. Lavaei, Optical rectification in quantum wells within different confinement and nonlinearity regimes. Superlattices Microstruct. **111**, 556–567 (2017)
48. Q. Yu, K. Guo, M. Hu, Z. Zhang, K. Li, D. Liu, Study on the optical rectification and second-harmonic generation with position-dependent mass in a quantum well. J. Phys. Chem. Solids **119**, 50–55 (2018)
49. C. Duque, M. Mora-Ramos, C. Duque, Hydrostatic pressure and electric field effects and nonlinear optical rectification of confined excitons in spherical quantum dots. Superlattices Microstruct. **49**, 264–268 (2011)
50. S. Dahiya, S. Lahon, R. Sharma, Effects of temperature and hydrostatic pressure on the optical rectification associated with the excitonic system in a semi-parabolic quantum dot. Phys. E: Low-dimensional Syst. Nanostruct. **118**, 113918 (2020)
51. S. Adachi, GaAs, AlAs, and $Al_xGa_{1-x}As$: material parameters for use in research and device applications. J. Appl. Phys. **58**(3), R1–R29 (1985)
52. Z.-Y. Deng, J.-K. Guo, T.-R. Lai, Impurity states in a spherical GaAs-$Ga_{1-x}Al_xAs$ quantum dot: effects of the spatial variation of dielectric screening. Phys. Rev. B **50**(8), 5736 (1994)
53. Y. Duan, X. Li, Spherical quantum dots in deng–fan–eckart potential confining potential: study of nonlinear optical rectification. Optik **265**, 169540 (2022)
54. O. Bayrak, I. Boztosun, H. Ciftci, Exact analytical solutions to the Kratzer potential by the asymptotic iteration method. Int. J. Quantum Chem. **107**(3), 540–544 (2007)
55. D. Hayrapetyan, E. Kazaryan, L. Petrosyan, H. Sarkisyan, Core/shell/shell spherical quantum dot with Kratzer confining potential: Impurity states and electrostatic multipoles. Phys. E: Low-dimensional Syst. Nanostruct. **66**, 7–12 (2015)
56. X. Liu, L. Zou, C. Liu, Z.-H. Zhang, J.-H. Yuan, The nonlinear optical rectifica- tion and second harmonic generation in asymmetrical Gaussian potential quantum well: effects of hydrostatic pressure, temperature and magnetic field. Opt. Mater. **53**, 218–223 (2016)
57. R.W. Boyd, Nonlinear Optics. (Academic Press, 2020)
58. R.-Z. Wang, K.-X. Guo, Z.-L. Liu, B. Chen, Y.-B. Zheng, Nonlinear optical rectification in asymmetric coupled quantum wells. Phys. Lett. A **373**, 795–798 (2009)

Hydrostatic Pressure and Temperature Effects on Nonlinear Optical Rectification in Tetrapod Quantum Dots

A. Ed-Dahmouny, A. Alkhaldi, M. Jaouane, H. M. Althib, R. Arraoui, A. Fakkahi, N. Zeiri, A. Sali, and C. A. Duque

1 Introduction

There are various types of semiconductor nanostructures, including tetrapods, octapods, and hyperbranched nanoparticles. Combining the electronic and optical properties of quantum dots with the properties of tetrapods exhibiting low lasing thresholds, dual electroluminescence, moderately good efficiency in solar cells, carriers, separation, directing, and reducing internal strain, which opens new and promising applications [2–6]. The behavior of quantum dots under electric and magnetic fields has been studied recently. GaAs and CdTe/CdS tetrapod quantum dots (TQDs) are highly dependent on geometry, size, and field intensity [1]. Semiconductor quantum dots can mediate electric-dipole, magnetic-dipole, and electric-quadrupole transitions on a single digital resonance, allowing them to probe electric and magnetic fields simultaneously [7]. The edge states of rectangular phosphorene quantum dots

A. Ed-Dahmouny (✉)
LaRSI Laboratory, Faculty of Sciences and Technology, Sidi Mohamed Ben Abdellah University, Fez, Morocco
e-mail: ayoub.eddahmouny@usmba.ac.ma

A. Alkhaldi · H. M. Althib
Department of Physics, College of Science, Imam Abdulrahman Bin Faisal University, Dammam, Saudi Arabia

M. Jaouane · R. Arraoui · A. Fakkahi · A. Sali
LPS Laboratory, Faculty of Science Dhar El Mahraz, Sidi Mohamed Ben Abdellah University, Fez, Morocco

N. Zeiri
Laboratoire de la Matière Condensée et des Nanosciences (LMCN), Département de Physique, Faculté des Sciences de Monastir, Monastir, Tunisia

C. A. Duque
Grupo de Materia Condensada-UdeA, Instituto de Física, Facultad de Ciencias Exactas y Naturales, Universidad de Antioquia UdeA, Medellín, Colombia

R. En-nadir (ed.), *Quantum Systems at the Nanoscale*, Lecture Notes in Nanoscale Science and Technology 45, https://doi.org/10.1007/978-3-032-06645-9_5

lead to three types of optical transitions that respond differently to applied fields [8]. Moreover, Mn-doped CdSe quantum dots exhibit light-induced spontaneous magnetization due to quantum confinement effects, resulting in long exciton lifetimes and strong exchange fields [9]. For high-capacity multiplexed detection applications, CdSe/CdS TQDs with a large Stokes shift have been developed to eliminate energy transfer through quantum dot (QD)-encoded microbeads [3]. The theoretical study investigates the influence of geometrical modifications and external electric and magnetic fields on confined electrons' energy levels and wave functions in GaAs and CdTe/CdS tetrapod-shaped nanostructures. The results show that for GaAs tetrapods, energy progressively reduces at all levels as the volume of the tetrapod increases through modification [1]. At 600 and 800 °C, ZnO tetrapod nanostructures doped with Mn by diffusion method showed increased size without any new X-ray diffraction peaks, and ferromagnetism with a Curie temperature around 50 K and large coercive fields [10, 11]. When the particle volume is increased in CdSe/CdS heterostructures, both the CdSe core and CdS arms exhibit efficient dual emission, modulated by excitation influence. Shape affects emission efficiency only marginally, but particle volume is the key to controlling Auger processes and understanding multiexciton emission, which makes these structures suitable for optical applications [12]. The size-dependent optical properties of tetrapod-shaped CdTe nanocrystals have been investigated [13]. The results indicate that the linear absorption spectra, photoluminescence excitation, and fluorescence experiments at room and low temperatures exhibit features unique to their complex geometry and reveal a clear correlation between oscillator strength and nanocrystal shape. The quasiparticle equation can be derived from classical electrostatics under suitable approximations and is essential for accurately describing the optical properties and addition energy variations in nonconvex nanosystems like tetrapods [14]. The addition energies vary due to spatial localization of quasiparticle states for tetrapods, unlike the constant behavior seen in spherical quantum dots. The temperature-dependent absorption and fluorescence spectra of colloidal CdSe nanocrystals study of CdSe nanocrystals revealed that while fluorescence peak shifts are shape-independent, spectral broadening differs significantly—especially in rods—indicating distinct broadening mechanisms, with tetrapods resembling dots because of an exciton state centered around the core [15]. High-frequency electron paramagnetic resonance (HF-EPR) spectroscopy suggests that lattice mismatch strain is relieved by interfacial alloying and misfit dislocations at the CdSe/ZnS core—arm interface in TQDs [16]. ZnO tetrapods showed quantum efficiency of 2–4%, with the optimally grown sample Z3 showing the highest efficiency, about 4.3%, and longest lifetime; this confirms that precision temperature control enhances quality and radiative recombination dominates relaxation [17]. When the particle volume is increased in CdSe/CdS heterostructures, both the CdSe core and CdS arms exhibit efficient dual emission, modulated by excitation influence. Through quantum dot seeds with accessible (111) planes, nanoscale tetrapods were synthesized, enabling tunable cores and arms [18]. However, cation exchange during growth often resulted in alloyed cores, except for Cd_3P_2/CdS tetrapods, which showed clean interfaces and rare near-infrared emission. The shape of single-crystalline nanotetrapods can be tuned to control exciton behavior, enabling a transition from 3D confinement to 2D confinement, multiphoton emission, and enhanced quantum effects like the Stark and memory effects [19]. A theoretical study explores how shape anisotropy and structure size in tetrapod nanocrystals, specifically variations in arm length or width, affect their electronic and optical properties [20]. These variations have a significant impact on light emissions, carrier lifetimes, wavefunction localization, radiative lifetimes, emission polarization, and Auger processes. The experiments show that CdSe/CdS tetrapods under mechanical stress exhibit changes in optical emission, making them accurate strain gauges at nanonewtons [21, 22]. Optomechanical devices and biological force sensors can benefit from their unique geometry and tunable structural and electronic properties. Spectrally distinct, dual-wavelength electroluminescence was observed in CdSe/CdS tetrapod-based LEDs under low bias, with emission originating from both the CdSe core and CdS arms [23]. Unlike nanorods, which exhibit a faster core localization and no dual emission, CdS arms exhibit long-lived excitons caused by surface trap states. The tetrapod quantum wells (QWs) were synthesized using a one-pot colloidal method using CdSe cores and gradient alloyed CdSeS arms [24]. This material exhibits dual photoluminescence peaks arising from separate transitions between the core and QWs, with tunable emission properties under variable excitation powers for light-emitting applications. By optimizing conditions such as injection temperature, precursor ratio, and growth temperature for synthesizing CdSe quantum dots and CdSe–ZnSe nano-tetrapods, it yields high-quality nanostructures with improved optical properties and is adaptable to industrial-scale production [25]. Tetrapod CdSe/CdS quantum dots, known for their anisotropic optical properties and large absorption cross-section, are typically synthesized at high temperatures using phosphonic acid ligands [26, 27]. In this study, amine ligands are used in conjunction with microfluidic control at low-temperature (120 °C) to simplify the growth mechanism for light-emitting materials and enable more sustainable production. The extant literature on the fluctuation of the NOR coefficient encompasses a range of theoretical investigations into the NOR variation of charge carriers confined within nanostructures of diverse geometries, including double QDs [28], spherical QD [29], cylindrical QD [30], and QWs [31]. This research endeavors to elucidate the impact of hydrostatic pressure and temperature on the transition energies associated with the two lowest-lying transitions of a donor impurity embedded in a GaAs TQD defined by an AlGaAs barrier, as well as the consequent effects of these perturbations on the NOR coefficients. Employing the effective mass and parabolic band approximations, numerical computations were executed via the FEM approach. The subsequent exposition is structured as follows: Sect. 2 details the theoretical framework employed; Sect. 3 presents and analyzes the numerical findings; and Sect. 4 provides a concise summary of the study.

2 Theoretical Framework

2.1 The Schrödinger Equation: Incorporating Simultaneous Pressure and Temperature Effects

In this chapter, we model a single electron interacting with a donor-impurity confined within a GaAs/Al$_\eta$Ga$_{1-\eta}$As TQDs subjected to hydrostatic pressure and temperature. The origin of our reference system is located at the geometric center of the TQD. Figure 1 provides a schematic representation of the nanostructure, where R_c and R_s denote the inner radius of the core and the outer radius of the shell, respectively, and H represents the height of each of the four arms. These dimensions are used throughout this work. The tetrapod structure was generated using COMSOL-Software [32] by intersecting four cylinders, ensuring that the core material shares the same crystal structure as the arms.

Within the effective mass approximation (EMA) and assuming parabolic bands, the Schrödinger equation reduces to the following differential equation in spherical coordinates:

$$H = -\frac{\hbar^2}{2}\nabla \cdot \left(\frac{1}{m^*_{C,S}(T,P)}\nabla\right) - \frac{\mathrm{k}\cdot e^2}{\varepsilon_{C,S}(T,P)|\vec{r}-\vec{r_i}|} + V_{Conf}(T,P,\vec{r}) \qquad (1)$$

The second term of the Hamiltonian models the interaction between an electron and a donor impurity located at spatial coordinate $\vec{r_i} = (0, 0, z_i \cdot \vec{e_z})$. Here, $k = 0$ indicates the absence and $k = 1$ the presence of the impurity. The z-axis of the reference frame $R(O, \vec{e_x}, \vec{e_y}, \vec{e_z})$ is aligned with the axis of revolution of a tetrapod arm.

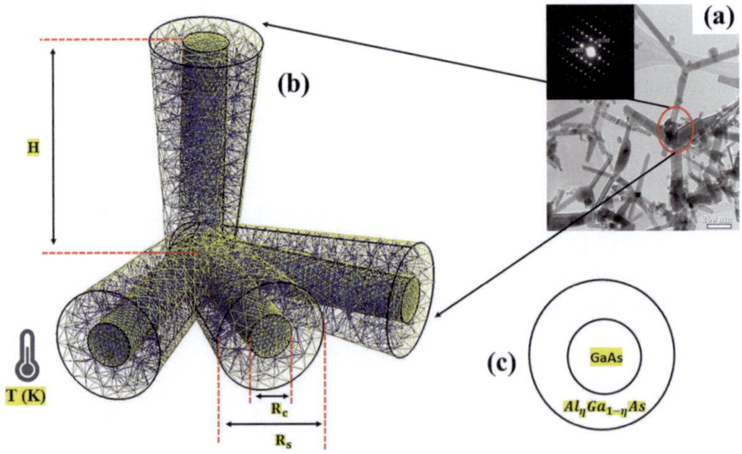

Fig. 1 **a** Transmission electron microscopy (TEM) image of a tetrapod nanostructure [4], **b** schematic diagram of a GaAs/Al$_\eta$Ga$_{1-\eta}$As TQD and **c** and arm cross-section zoom [33]

The term $V_{Conf}(T, P, \vec{r})$, the third term in the equation, denotes the hydrostatic pressure and temperature-dependent confinement potential within a GaAs/$Al_\eta Ga_{1-\eta}As$ TQD. Its mathematical expression is as follows:

$$V_{Conf}(T, P, \vec{r}) = \begin{cases} 0 & \text{in GaAs} \\ V_0(T, P, \eta) & \text{in } Al_\eta Ga_{1-\eta}As \\ \infty & \text{otherwise} \end{cases} \quad (2)$$

$V_0(T, P, \eta)$ represents the finite barrier height, a function of temperature, hydrostatic pressure, and the mole fraction of aluminum in the barrier layer, and is defined by [34]:

$$V_0(T, P, \eta) = 0.65 \times \left[\Delta E_g^\Gamma(\eta) + A(\eta) \times P + B(\eta) \times T\right] \quad (3)$$

Here, $\Delta E_g^\Gamma(\eta)$ represents the energy gap difference between the dot material GaAs and the barrier material $Al_\eta Ga_{1-\eta}As$ at the Γ-point, independent of temperature and pressure, and is given by the formula: $\Delta E_g^\Gamma(\eta) = [1.155 \times \eta + 0.37 \times \eta^2]$ (eV). The coefficients $A(\eta)$ and $B(\eta)$ quantify the contributions of pressure and temperature, respectively, to the bandgap shift and are expressed as follows: $A(\eta) = -1.3 \cdot \eta \times 10^{-3}$ (eV/Kbar) and $B(\eta) = -1.115 \cdot \eta \times 10^{-4}$ (eV/K).

The application of external hydrostatic pressure primarily influences the dimensions of the TQD nanostructure, specifically the core and shell TQD dimensions (cone heights H and outer R_s and inner R_c arm radii), which are expressed as functions of pressure by [35]:

$$\begin{cases} H(P) = H(P = 0) \cdot [1 - (S_{11} + 2S_{12}) \cdot P] \\ R_c(P) = R_c(P = 0) \cdot [1 - (S_{11} + 2S_{12}) \cdot P] \\ R_s(P) = R_s(P = 0) \cdot [1 - (S_{11} + 2S_{12}) \cdot P] \end{cases} \quad (4)$$

In addition to hydrostatic pressure, the application of temperature, both independently and concurrently with pressure, influences the effective masses of charge carriers in materials, depending on their degrees of freedom and dielectric constants, and is given by the following expression. Therefore, the effective masses of electrons confined within the GaAs core or $Al_\eta Ga_{1-\eta}As$ shell, which are dependent on hydrostatic pressure and temperature, can be determined using the following expression [35]:

$$\begin{cases} m_C^*(T, P, \eta) = m_0 \cdot \left[1 + 7.15\left(\frac{2}{E_{g,\Gamma}^{(T,P)}} + \frac{1}{E_{g,\Gamma}^{(T,P)} + \Delta_0}\right)\right]^{-1} \\ m_S^*(T, P, \eta) = m_0 \cdot \left(m_C^*(T, P, \eta) + 0.0835 \cdot \eta\right) \end{cases} \quad (5)$$

Here, m_0 represents the free electron mass, and $\Delta_0 = 0.341$ eV is the spin-orbit splitting of the GaAs valence band. In the preceding expression, the pressure (P) is in kbar and the temperature (T) is in K. Throughout this paper, the aluminum concentration η in the barrier layer is fixed at 0.3. The temperature and hydrostatic pressure dependence of the GaAs energy gap at the Γ-point, expressed in eV, is given by the following formula [35]:

$$E_{g,\Gamma}^{(T,P)} = E_{g,\Gamma}^{(T,0\,\mathrm{kbar})} + 10.7 \times 10^{-3} \cdot P \tag{6}$$

where $E_{g,\Gamma}^{(T,0\,\mathrm{kbar})}$ [eV] represents the energy gap at zero hydrostatic pressure, defined by [35]:

$$E_{g,\Gamma}^{(T,0\,\mathrm{kbar})} = 1.519 - \frac{5.405 \times 10^{-4} \times T^2}{T + 204} \tag{7}$$

The static dielectric constants of the core and shell materials, $\varepsilon_C(T, P)$ and $\varepsilon_S(T, P, \eta)$, which are dependent on hydrostatic pressure and temperature, are given by [36]:

$$\varepsilon_C(T, P) = \begin{cases} 12.74 \times e^{-1.67\times 10^{-3}\times P}\, e^{9.4\times 10^{-5}\times(T-75.6)} & \text{for T} \leq 200\text{ K} \\ 13.18 \times e^{-1.73\times 10^{-3}\times P}\, e^{20.4\times 10^{-5}\times(T-300)} & \text{for T} > 200\text{ K} \end{cases} \tag{8}$$

and

$$\varepsilon_S(T, P) = \varepsilon_C(T, P) - 3.12 \cdot \eta \tag{9}$$

2.2 *Nonlinear Optical Rectification via Finite Element Methods*

This study investigates NOR coefficient in a GaAs/$Al_\eta Ga_{1-\eta}As$ TQD with an Al concentration of 0.3. We examine the combined influence of hydrostatic pressure and temperature on the NOR coefficient, both in the presence and absence of a donor impurity. For impurity cases, we analyze its impact at various positions along the tetrapod's symmetry axis (Oz). Utilizing the density matrix formalism, we model NOR arising from sub-band transitions within the core-shell structure, driven by a polarized monochromatic electric field from a laser, as described in [37]:

$$\hat{E}(t) = \hat{E} \cdot e^{-i\omega t} + \hat{E} \cdot e^{i\omega t} \tag{10}$$

A strong electromagnetic field disturbs the equilibrium charge distribution within a medium, leading to modifications in its electronic structure and optical properties. The resulting change in macroscopic polarization varies with the applied electric field strength. By employing a third-order series expansion to approximate susceptibility, the electronic polarization of core-shell nanostructures can be expressed as given in [30]:

$$\hat{P}(t) = \varepsilon_r \left[\chi^{(1)} \hat{E}(t) + \chi^{(2)} \hat{E}(t)^2 + \chi^{(3)} \hat{E}(t)^3 + \cdots \right] \\ = \hat{P}^{(1)}(t) + \hat{P}^{(2)}(t) + \hat{P}^{(3)}(t) + \cdots \quad (11)$$

The coefficients $\chi^{(1)}$, $\chi^{(2)}$, and $\chi^{(3)}$ represent the first-, second-, and third-order nonlinear optical susceptibilities, respectively. Thus, the polarization can be rewritten as [30]:

$$\hat{P}(t) = \varepsilon_r \chi_{\omega}^{(1)} \hat{E} e^{i\omega t} + \varepsilon_r \chi_0^{(2)} \hat{E}^2 + \varepsilon_r \chi_{2\omega}^{(2)} \hat{E}^2 e^{2i\omega t} + \varepsilon_r \chi_{\omega}^{(3)} \hat{E}^3 e^{i\omega t} + \varepsilon_r \chi_{3\omega}^{(3)} \hat{E}^3 e^{3i\omega t} \quad (12)$$

Our investigation centers on the second-order nonlinear polarization within this nanomaterial, determined by formula [30]:

$$\hat{P}^{(2)}(t) = 2\varepsilon_r \chi_0^{(2)} \hat{E}^2 + \varepsilon_r \chi_{2\omega}^{(2)} \hat{E}^2 e^{-2i\omega t} + cc \quad (13)$$

ε_r denotes the permittivity of both the semiconductor core and shell. The first term, $\chi_0^{(2)}$, represents NOR coefficient, a second-order polarization process that generates a static electric field within the nonlinear crystal. The final expression for the NOR coefficient, derived using the density matrix formalism, is provided by Rosencher and Bois [37] and further detailed in [38]:

$$\chi_0^{(2)} = \frac{4e^3\sigma_s}{\varepsilon_0 \hbar^2} |M_{if}^z|^2 \delta_{if} \left(\frac{\omega_{if}^2 \left(1 + \frac{\Gamma_2}{\Gamma_1}\right) + \left(\omega^2 + \Gamma_2^2\right)\left(1 - \frac{\Gamma_2}{\Gamma_1}\right)}{\left((\omega_{if} - \omega)^2 + \Gamma_2^2\right)\left((\omega_{if} + \omega)^2 + \Gamma_2^2\right)} \right) \quad (14)$$

where $M_{if} = \langle \Psi_i | z | \Psi_f \rangle$ represents the off-diagonal matrix element $\delta_{if} = | \langle \Psi_i | z | \Psi_i \rangle - \langle \Psi_f | z | \Psi_f \rangle |$ is the difference between the corresponding matrix elements of the two transition states. $\omega_{if} = \frac{E_f - E_i}{\hbar}$ signifies the Bohr frequency under resonant conditions, being close to the incident radiation frequency ω. The diagonal and off-diagonal relaxation rates are given by Γ_1 and Γ_2, respectively, with σ_s representing the system's electron density.

The quantized energy spectrum of charge carriers confined within $GaAs/Al_\eta Ga_{1-\eta} As$ core/shell TQD is governed by partial differential equations (PDEs), specifically the time-independent Schrödinger equation. However, an analytical solution to this eigenvalue problem becomes intractable in the presence of a donor impurity due to introducing a non-separable Coulombic interaction that couples the spatial degrees of freedom of the electron and the impurity center. Consequently, approximate solutions require the application of numerical methodologies. Among the repertoire of available numerical techniques, the FEM technique exhibits a notable aptitude for yielding accurate solutions with minimized numerical uncertainty, thereby providing a robust representation of the underlying physical phenomena. The efficacy of FEM is evidenced by its widespread adoption in various domains within scientific research and engineering [39]. The fundamental principle of FEM involves the discretization of the computational domain into a tessellation of elementary (finite) elements or a mesh, a procedure termed meshing. Within this discretized framework, if we divide the structure into $n_{elements}$, the electronic wave function, which encapsulates the quantum state and associated physical observables of the electron, can then be approximated as follows:

$$\Psi_i = \sum_{j=1}^{n_{elements}} A_j \varphi_j \tag{15}$$

where φ_j denotes the base functions and A_j denotes the coefficients of these base functions.

We can adapt the Hamiltonian described in Eq. (1) and formulate the stationary Schrödinger equation within the framework of the FEM approach. This involves multiplying both sides of the Schrödinger equation by a test function ϕ_k and integrating over the spatial domain of the nanostructure: $\Omega = (\mathrm{GaAs}) \cup (\mathrm{Al}_\eta \mathrm{Ga}_{1-\eta}\mathrm{As})$:

$$\sum_{j,k=1}^{n_{\text{elements}}} \int_\Omega \phi_k \left(-\frac{\hbar^2}{2} \nabla \cdot \left(\frac{1}{m^*_{C,S}(T,P)} \nabla \right) - \frac{\mathrm{k} \cdot e^2}{\varepsilon_{C,S}(T,P)|\vec{r} - \vec{r}_i|} \right.$$
$$\left. + V_{Conf}(T,P,\vec{r}) \right) \varphi_i(\vec{r}_j) d\vec{r}_j = \sum_{j,k=1}^{n_{\text{element}}} \int_\Omega E_i \phi_k \varphi_i(\vec{r}_j) d\vec{r}_j \tag{16}$$

We separate the various terms within this integral. Then, applying integration by parts, the equation can be expressed as follows:

$$\sum_{j,k=1}^{n_{\text{elements}}} \left\{ \int_\Omega \frac{\hbar^2}{2m^*} \nabla \phi_k \cdot \nabla \varphi_i(\vec{r}_j) d\vec{r}_j - \left[\frac{\hbar^2}{2m^*} \phi_k \frac{d\varphi_i(\vec{r}_j)}{dr} \right] \right.$$
$$\left. + \int_\Omega \left(-\frac{\mathrm{k} \cdot e^2}{\varepsilon_{C,S}(T,P)|\vec{r} - \vec{r}_i|} + V_{Conf}(T,P,\vec{r}) \right) \phi_k \varphi_i(\vec{r}_j) d\vec{r}_j \right\}$$
$$= \sum_{j,k=1}^{n_{\text{elements}}} \int_\Omega E_i \phi_k \varphi_i(\vec{r}_j) d\vec{r}_j \tag{17}$$

To determine the eigenvalues and eigenstates of a Coulombic interacting electron-impurity pair confined in a three-dimensional tetrapod-like core-shell nanostructure, we utilized the FEM approach within the COMSOL Multiphysics environment [32]. This numerical solution of the Schrödinger equation (Eq. 17) was achieved with high accuracy by employing an optimized mesh and carefully implementing the Dirichlet and Neumann boundary conditions at the core-shell interfaces [40]. Valuable insights into these properties, particularly the NOR behavior presented in this paper, are derived from the resulting eigenvalues and eigenfunctions.

3 Results and Discussion

Figure 2 presents a comprehensive analysis of the effects of hydrostatic pressure P on the transition energies for two distinct transitions: E_{12} (ground to first excited state) and E_{23} (first to second excited state) in GaAs/Al$_\eta$Ga$_{1-\eta}$As TQD structure. The investigation considers two cases: the absence of donor impurities (k = 0) and the presence of donor impurities (k = 1), considering also two impurity locations at $z_0 = 0$ nm (the center of the TQDs) and at $z_0 = 20$ nm (far from the center). The results show the impact of increasing pressure (0–40 kbar) with fixed temperature at T = 0 K, as illustrated in Fig. 2. The increasing of hydrostatic pressure leads to a notable shift in both transition energies: E_{12} and E_{23}, which can be attributed to

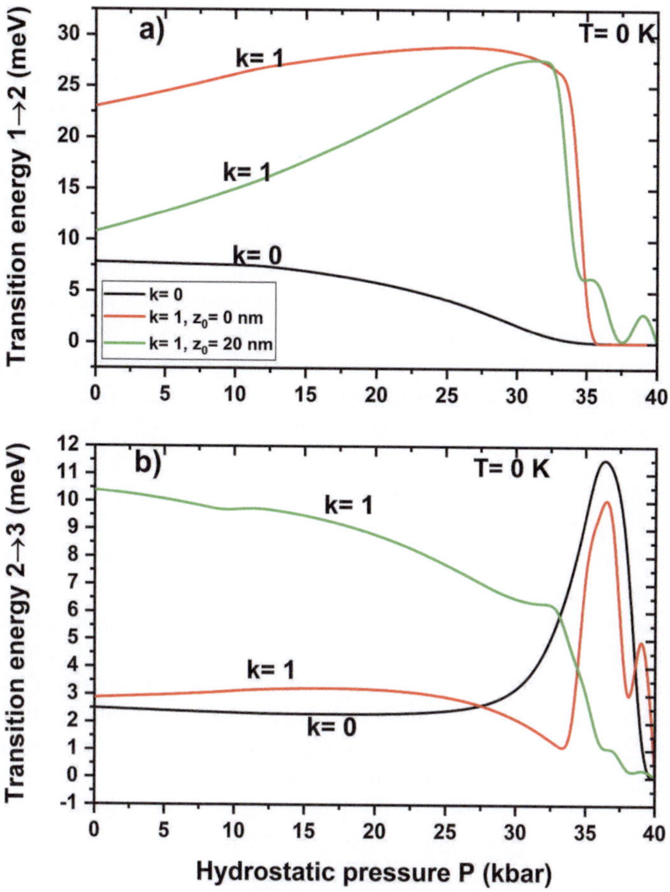

Fig. 2 Effect of hydrostatic pressure on E_{12} (**a**) and E_{23} (**b**) transition energies: comparison of impurity positions: $z_0 = 0$ nm and 20 nm and in the case of absence of impurity (k = 0). The simulations were conducted at zero temperature with a fixed system size

changes in the confinement potential and the overall band structure of the TQDs. These changes are due to the stress acting on the TQDs structure and the pressure that modifies material parameters. It is clear from Eq. (4) that under hydrostatic pressure, the cone inner radius R_c, outer radius R_s, and arm height H contract linearly, leading to a more substantial quantum confinement potential. This also occurs because energy levels are inversely proportional to the square of the confinement length $E_n \propto \frac{1}{L^2}$, causing a monotonic rise in the transition energies. In addition, the Coulomb interaction between the electron and donor amplifies the Stark shift of energy levels and slightly increases the carriers' effective mass with pressure, as shown in Eq. (5). It obviously the effect of donor impurities (k = 1) significantly increases E_{12} and E_{23} compared to the free-impurity case (k = 0). This behavior is due to the Coulomb attraction that reduces the ground state energy E_1 compared to higher states E_2, E_3, leading to an increase in the band gap energy. In addition to the presence or absence impurities, another factor plays an important role the spatial location of the impurities: central impurities ($z_0 = 0$ nm) maximize the overlap of the wavefunction with the Coulomb potential, yielding the largest transition energies, while non-central impurities ($z_0 = 20$ nm) cause asymmetric perturbation of the system, reducing the interaction strength. The greater sensitivity of E_{23} to impurity displacement compared to E_{12} indicates that higher energy states ($n = 3$) exhibit more spatially extended wavefunctions, making them sensitive to local changes in impurity potentials. At higher pressures, the observed behavior in Fig. 2 continues to increase until reach ($\sim$ 30 kbar) the transition energy E_{12} drops to zero, and the behavior of a centered impurity becomes similar to a free impurity, while E_{23} continues to increase. This is due to the fact that beyond 30 kbar, the impurity's effect becomes negligible, meaning that both impurity-bound and free-electron cases experience similar energy shifts. Additionally, deviations may appear due to nonlinear lattice distortion or phase transitions in the GaAs/$Al_{\eta}Ga_{1-\eta}As$ heterostructure. This behavior is consistent with deformation potential theory, where strain-induced bandgap modifications dominate over effects such as carrier-phonon coupling. The tunability of transition energies via pressure and impurity engineering shows the potential of TQDs in adaptive optoelectronic applications, such as pressure-sensitive photodetectors or impurity-tunable single-photon emitters.

Figure 3 investigates the temperature dependence of E_{12} (a–c) and E_{23} (d–f) for impurity-free (k = 0) and impurity-presence ($z_0 = 0$ nm, $z_0 = 20$ nm) in TQDs. The results show that the transition energy increases with temperature, except for the E_{12} in the presence of impurities case, as observed in Fig. 3b, c. The transition energies exhibit minimal variation at low temperatures, reflecting the relatively stable lattice interactions and confinement potential. However, as the temperature rises, the increased thermal excitation of the lattice leads to shifts in the energy levels due to thermal expansion. For impurity-free TQDs, increasing temperature weakens the confinement potential, allowing higher-energy states to shift upward faster than lower-energy states, leading to an overall increase in transition energy. This effect dominates in Fig. 3a, d, e, f, where all transitions show an upward increase with temperature. The presence of impurities introduces additional complexity to this behavior. The results show that in the case of E_{12} with an impurity, shown in Fig. 3b,

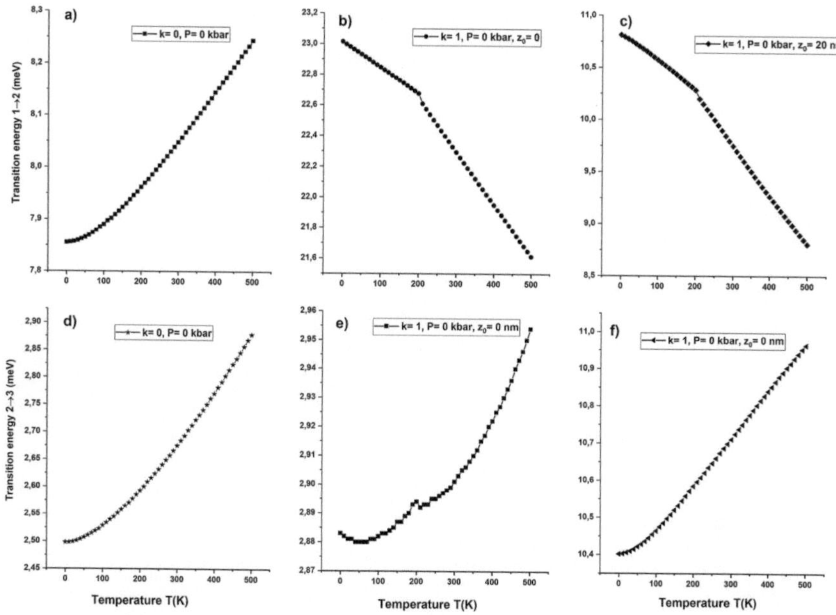

Fig. 3 Effect of temperature on E_{12} (**a–c**) and E_{23} (**d–f**) transition energies in the absence of an impurity (k = 0) and in its presence at two different positions: $z_0 = 0$ nm and 20 nm. The simulations were conducted at zero temperature with a fixed system size

c, a different behavior was observed. When an impurity exists, the ground state (E_1) is strongly localized due to the Coulomb attraction between the electron and the impurity. As temperature increases, thermal energy allows the electron to explore higher-energy states, which weakens this Coulomb interaction. As a result, the ground state energy (E_1) increases, but at a rate slower than the first excited state (E_2). This causes the energy difference to decrease, leading to the observed drop in transition energy for these specific cases. The dependence of the dielectric constant (Eq. 8) on temperature also weakens Coulomb interactions, especially above 200 K, where the value ε_C increases significantly. The nonlinear decrease in transition energies above 150 K is consistent with the dominance of longitudinal optical (LO) phonon scattering, as shown by the T^3-dependent electron-phonon scattering rate. These results shed light on the balance between thermal excitation, impurity effects, and quantum confinement.

Figures 4, 5, and 6 illustrate the NOR coefficient for transitions $1 \rightarrow 2$ and $2 \rightarrow 3$ as a function of impurity position, hydrostatic pressure, and temperature, respectively. Figure 4 analysis NOR as function of impurity position z_0 (0, 20 and 40 nm) fixed at T = 0 K and P = 0 kbar. The free impurity case exhibits a strong NOR coefficient. The behavior of NOR depends on several factors, including wavefunction symmetry, transition dipole moments, and impurity-induced modifications in the confinement potential. A centered impurity ($z_0 = 0$ nm) slightly modifies the wavefunctions, but

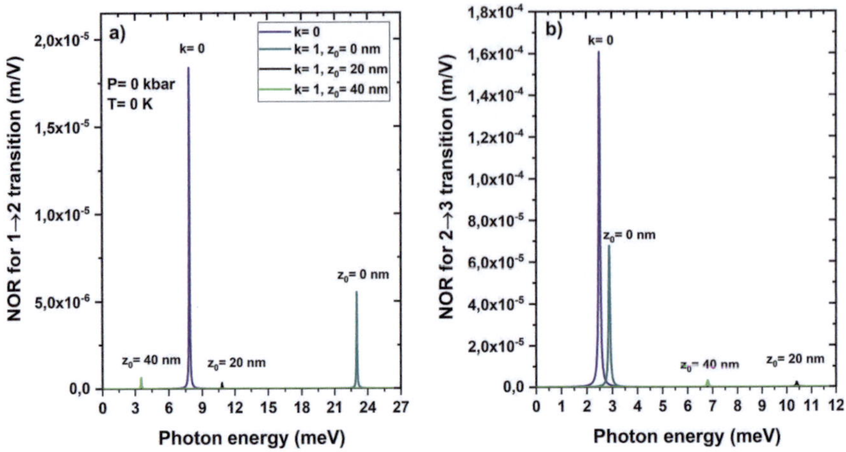

Fig. 4 The effect of impurity position z_0 on the NOR coefficient of the $1 \rightarrow 2$ and $2 \rightarrow 3$ transitions, simulated at zero temperature and pressure with a fixed system size

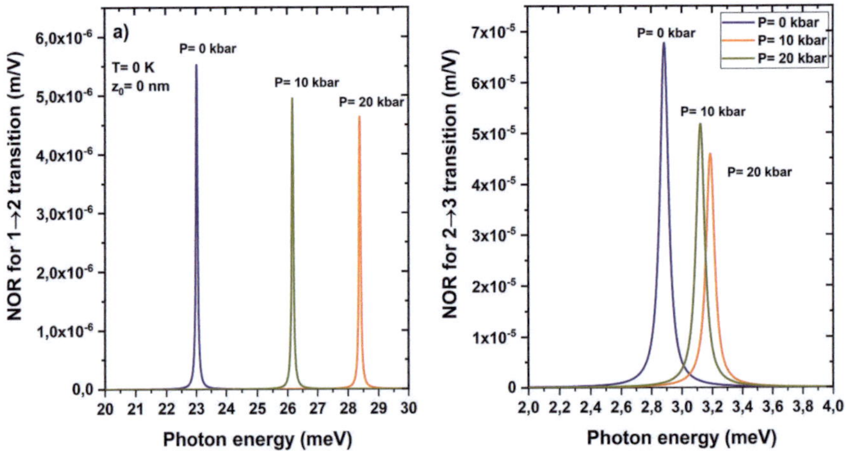

Fig. 5 The NOR coefficient for the $1 \rightarrow 2$ and $2 \rightarrow 3$ transitions of a confined on-center impurity in TQD was simulated at zero temperature with a fixed system size. The variation with incident photon energy was investigated in the absence of hydrostatic pressure and at two hydrostatic pressures: 10 and 20 kbar

under certain conditions, shows maximum wavefunction overlap optimizes dipole moments (μ_{12}, μ_{23}), which can enhance NOR due to additional polarization effects induced by Coulomb interaction. On the other hand, if impurity is far from the center, it breaks system symmetry significantly, reducing wavefunction overlap and weakening NOR. The $2 \rightarrow 3$ transition exhibits greater positional impurity sensitivity than $1 \rightarrow 2$, reflecting the delocalized nature of higher energy states, and the dipole

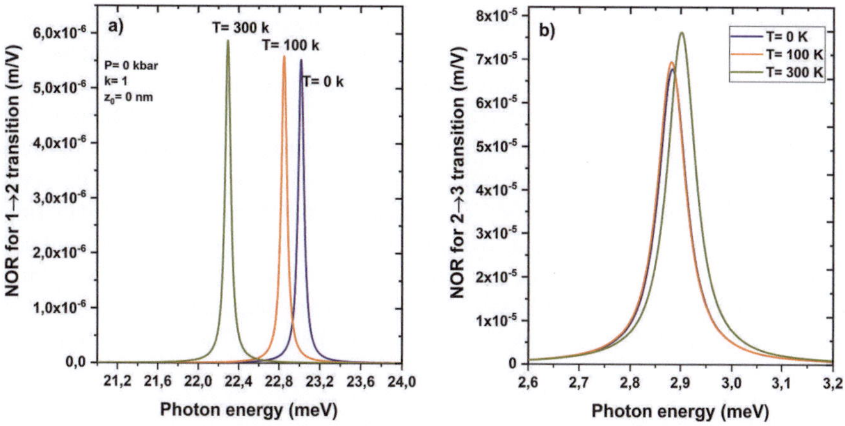

Fig. 6 The temperature dependence of the NOR coefficient was studied for the $1 \rightarrow 2$ and $2 \rightarrow 3$ transitions of an on-center impurity in a TQD. Simulations were performed at zero pressure with a fixed system size, as a function of incident photon energy, at temperatures of 0, 100 and 300 K

moment variations due to impurity shifts influence the NOR coefficient more strongly for $2 \rightarrow 3$ than $1 \rightarrow 2$.

Moving to Fig. 5, which presents the variation of the NOR coefficient with incident photon energy for a confined on-center impurity in the TQDs for $1 \rightarrow 2$ and $2 \rightarrow 3$ transitions, analyzed under different hydrostatic pressure conditions (0, 10, and 20 kbar) with a fixed system size at zero temperature. The results demonstrate that the NOR coefficient exhibits a pronounced dependence on incident photon energy and applied pressure, indicating the complex interplay between these parameters. At zero pressure (P = 0 kbar), the TQD has a relatively larger size and weaker confinement, leading to an enhanced wavefunction overlap between quantum states. This enhances transition dipole moments, which increases NOR. As pressure increases, the overall structure of the TQD crystal lattice changes, leading to a change in the localization of electronic states. Increasing pressure shifts the NOR resonance peaks due to strain-induced bandgap widening $E_g \propto P$ (Eq. 6), which shifts intersubband transitions to higher energies. As energy levels shift, the resonance conditions that lead to electron-photon interactions may not efficiently match the energies of the incoming photons. In addition, high pressure can enhance scattering effects and localization of charge carriers, leading to electronic disorder. All these phenomena reduce the coherence of the electronic states and diminish the efficiency of the nonlinear optical processes, resulting in a reduction of the NOR coefficient. Overall, the decrease in NOR with increasing pressure reflects the complex interaction between structural changes and electronic properties within the quantum dot system. The $2 \rightarrow 3$ transition shows a more significant pressure-induced shift than $1 \rightarrow 2$, due to its stronger sensitivity to confinement effects as seen in Fig. 5. The analysis suggests that tuning hydrostatic pressure can be utilized to optimize the NOR of the TQD, which is particularly relevant for applications in frequency conversion and optical switching.

Turning to Fig. 6, it explores the dependence of the NOR coefficient on the variations of the temperature of the center impurity TQD structure for transitions $1 \to 2$ and $2 \to 3$. The results demonstrate how increasing temperature from 0 to 300 K, influences the NOR coefficient. As the temperature increases, the NOR coefficient typically shows a noticeable enhancement owing to thermal excitation and improved electron mobility. At higher temperatures, the electronic states in the TQD are more stable, leading to higher NOR values. This stability results from the increased thermal excitation, which allows electrons to occupy energy states more effectively and participate in nonlinear optical processes. The $2 \to 3$ transition exhibits greater thermal suppression than $1 \to 2$ since higher-energy states are generally more delocalized and are more affected by thermal fluctuations and electron-phonon interactions.

4 Concluding Remarks

This study presents an innovative analysis of the electronic behavior and NOR coefficient of an electron confined within a core/shell TQDs, considering the combined influences of hydrostatic pressure and temperature. We investigated the two lowest energy transitions, both with and without the presence of a donor impurity. Our findings reveal that increasing hydrostatic pressure leads to a significant blue shift in both transition energies (E_{12} and E_{23}). Furthermore, the presence of a donor impurity ($k = 1$) noticeably increases these transition energies compared to the impurity-free case. Regarding temperature effects, the transition energies exhibit minimal variation at low temperatures, indicating the stability of lattice interactions and the confinement potential. Conversely, the NOR associated with the $2 \to 3$ transition demonstrates greater sensitivity to the positional impurity than the $1 \to 2$ transition. This heightened sensitivity likely arises from the more delocalized nature of higher energy states, where impurity-induced variations in the dipole moment exert a stronger influence on the NOR coefficient. The results further demonstrate a pronounced dependence of the NOR coefficient on incident photon energy and applied pressure, highlighting the intricate interplay between these parameters. Finally, increasing temperature generally enhances the NOR coefficient, likely due to thermal excitation and improved electron mobility.

Acknowledgements CAD is grateful to the Colombian Agencies: CODI-Universidad de Antioquia (Estrategia de Sostenibilidad de la Universidad de Antioquia and projects "Complejos excitónicos y propiedades de transporte en sistemas nanométricos de semiconductores con simetría axial basados en GaAs e InAs", and "Nanoestructuras semiconductoras con simetría axial basadas en InAs y GaAs para aplicaciones en electrónica ultra e híper rápida") and Facultad de Ciencias Exactas y Naturales-Universidad de Antioquia (CAD exclusive dedication project 2025–2026).

Data Availability The corresponding author can provide all files related to tables, figures, and code on request.

Declaration of Competing Interest The authors have no conflicts of interest to disclose.

References

1. R.V. Hahn, F. Mora-Rey, R.L. Restrepo, A.L. Morales, J. Montoya-Sánchez, G. Eramo, M.G. Barseghyan, A. Ed-Dahmouny, J.A. Vinasco, D.A. Duque, C.A. Duque, Electronic and optical properties of tetrapod quantum dots under applied electric and magnetic fields. Eur. Phys. J. Plus **139**, 1–9 (2024). https://doi.org/10.1140/epjp/s13360-024-05089-z
2. R.A. Glass, Solterra Renewable Technologies, Sustainable and cost effective solar technology made from tetrapod quantum dot solar cells. TechConnect Briefs **3**, 729–731 (2010)
3. W. Wu, X. Yu, M. Gao, S. Gull, L. Shen, W. Wang, L. Li, Y. Yin, W. Li, Precisely encoded barcodes using tetrapod CdSe/CdS quantum dots with a large Stokes shift for multiplexed detection. Adv. Funct. Mater. **30**, 1–9 (2020). https://doi.org/10.1002/adfm.201906707
4. H.M. Cheng, K.Y. Huang, K.M. Lee, P. Yu, S.C. Lin, J.H. Huang, C.G. Wu, J. Tang, High-efficiency cascade CdS/CdSe quantum dot-sensitized solar cells based on hierarchical tetrapod-like ZnO nanoparticles. Phys. Chem. Chem. Phys. **14**, 13539–13548 (2012). https://doi.org/10.1039/c2cp41760j
5. N. Mishra, V.G. Vasavi Dutt, M.P. Arciniegas, Recent progress on metal chalcogenide semiconductor tetrapod-shaped colloidal nanocrystals and their applications in optoelectronics. Chem. Mater. **31**, 9216–9242 (2019). https://doi.org/10.1021/acs.chemmater.8b05363
6. C.K. Gordon, L.D. Browne, S. Chan, M.W. Brett, C. Zemke-Smith, J. Hardy, M.B. Price, N.J.L.K. Davis, Heterostructured nanotetrapod luminophores for reabsorption elimination within luminescent solar concentrators. ACS Appl. Mater. Interfaces **15**, 17914–17921 (2023). https://doi.org/10.1021/acsami.3c01222
7. P. Tighineanu, M.L. Andersen, A.S. Sorensen, S. Stobbe, P. Lodahl, Probing electric and magnetic vacuum fluctuations with quantum dots. Phys. Rev. Lett. **113**, 43601 (2014). https://doi.org/10.1103/PhysRevLett.113.043601
8. L.L. Li, D. Moldovan, W. Xu, F.M. Peeters, Electric- and magnetic-field dependence of the electronic and optical properties of phosphorene quantum dots. Nanotechnology **28**, 85702 (2017). https://doi.org/10.1088/1361-6528/aa55e8
9. R. Beaulac, L. Schneider, P.I. Archer, G. Bacher, D.R. Gamelin, Light-induced spontaneous magnetization in doped colloidal quantum dots. Science **325**, 973–976 (2009). https://doi.org/10.1126/science.1174419
10. V.A.L. Roy, A.B. Djurišić, H. Liu, X.X. Zhang, Y.H. Leung, M.H. Xie, J. Gao, H.F. Lui, C. Surya, Magnetic properties of Mn doped ZnO tetrapod structures. Appl. Phys. Lett. **84**, 756–758 (2004). https://doi.org/10.1063/1.1645312
11. C.A. Duque, N. Porras-Montenegro, Z. Barticevic, M. Pacheco, L.E. Oliveira, Effects of applied magnetic fields and hydrostatic pressure on the optical transitions in self-assembled InAs/GaAs quantum dots. J. Phys. Condens. Matter **18**, 1877 (2006). https://doi.org/10.1088/0953-8984/18/6/005
12. A.A. Lutich, C. Mauser, E. Da Como, J. Huang, A. Vaneski, D.V. Talapin, A.L. Rogach, J. Feldmann, Multiexcitonic dual emission in CdSe/CdS tetrapods and nanorods. Nano Lett. **10**, 4646–4650 (2010). https://doi.org/10.1021/nl1028057
13. D. Tarì, M. De Giorgi, F.D. Sala, L. Carbone, R. Krahne, L. Manna, R. Cingolani, S. Kudera, W.J. Parak, Optical properties of tetrapod-shaped CdTe nanocrystals. Appl. Phys. Lett. **87**, 224101 (2005). https://doi.org/10.1063/1.2130727
14. Wang, Charging effects in a CdSe nanotetrapod. J. Phys. Chem. B **109**, 23330–23335 (2005). https://doi.org/10.1021/jp054349q
15. A. Al Salman, A. Tortschanoff, M.B. Mohamed, D. Tonti, F. Van Mourik, M. Chergui, Temperature effects on the spectral properties of colloidal CdSe nanodots, nanorods, and tetrapods. Appl. Phys. Lett. **90** (2007). https://doi.org/10.1063/1.2696687
16. W. Zheng, Z. Wang, J. van Tol, N.S. Dalal, G.F. Strouse, Alloy formation at the tetrapod core/arm interface. Nano Lett. **12**, 3132–3137 (2012). https://doi.org/10.1021/nl3010447
17. M.C. Tam, A.M.C. Ng, A.B. Djurišić, K.S. Wong, Correlation of quantum efficiency and photoluminescence lifetime of ZnO tetrapods grown at different temperatures. J. Appl. Phys. **112**, 23515 (2012). https://doi.org/10.1063/1.4737785

18. M.J. Enright, F.Y. Dou, S. Wu, E.J. Rabe, M. Monahan, M.R. Friedfeld, C.W. Schlenker, B.M. Cossairt, Seeded growth of nanoscale semiconductor tetrapods: generality and the role of cation exchange. Chem. Mater. **32**, 4774–4784 (2020). https://doi.org/10.1021/acs.chemmater.0c01407
19. T. Kim, Y. Kim, S. Park, K. Park, Z. Wang, S.H. Oh, S. Jeong, D. Kim, Shape-tuned multiphoton-emitting InP nanotetrapods. Adv. Mater. **34**, 2110665 (2022). https://doi.org/10.1002/adma.202110665
20. M. Califano, Y. Zhou, Effects of structural asymmetries in branched nanocrystals: implications for electron collection efficiency in single-nanotetrapod devices. ACS Appl. Nano Mater. **5**, 18986–18997 (2022). https://doi.org/10.1021/acsanm.2c04801
21. C.L. Choi, K.J. Koski, S. Sivasankar, A.P. Alivisatos, Strain-dependent photoluminescence behavior of CdSe/CdS nanocrystals with spherical, linear, and branched topologies. Nano Lett. **9**, 3544–3549 (2009). https://doi.org/10.1021/nl9017572
22. S. Castelletto, Z. Bodrog, A.P. Magyar, A. Gentle, A. Gali, I. Aharonovich, Quantum-confined single photon emission at room temperature from SiC tetrapods. Nanoscale **6**, 10027–10032 (2014). https://doi.org/10.1039/C4NR02307B
23. J.I. Wong, N. Mishra, G. Xing, M. Li, S. Chakrabortty, T.C. Sum, Y. Shi, Y. Chan, H.Y. Yang, Dual wavelength electroluminescence from CdSe/CdS tetrapods. ACS Nano **8**, 2873–2879 (2014). https://doi.org/10.1021/nn500030t
24. H. Lee, H.J. Song, M. Shim, C. Lee, Towards the commercialization of colloidal quantum dot solar cells: perspectives on device structures and manufacturing. Energy Environ. Sci. **13**, 404–431 (2020). https://doi.org/10.1039/c9ee03348c
25. B. Samuel, K.J. Adarsh, V.P. Nampoori, A. Mujeeb, Optimization of certain synthesis parameters for CdSe quantum dots and synthesis of CdSe–ZnSe nano-tetrapods through facile hot injection method. Phys. Chem. Math. **14**, 479–488 (2023). https://doi.org/10.17586/2220-8054-2023-14-4-479-488
26. W. Xing, S. Zhang, R. An, W. Bi, C. Geng, S. Xu, Low-temperature synthesis of tetrapod CdSe/CdS quantum dots through a microfluidic reactor. Nanoscale **13**, 19474–19483 (2021). https://doi.org/10.1039/D1NR04070G
27. Y. Yao, Q. Shen, C. Wu, C. Lu, J. Sheng, Y. Li, H. Yang, Preparation of Ag_3PO_4 tetrapods anchored to nitrogen-doped carbon quantum dots for enhanced photocatalytic performance. New J. Chem. **47**, 10497–10505 (2023). https://doi.org/10.1039/D3NJ01042B
28. R. Arraoui, K. El-Bakkari, K. Limame, A. Ed-Dahmouny, M. Jaouane, A. Fakkahi, H. Azmi, A. Sali, Pressure and temperature influences on the nonlinear optical rectification of an impurity in a symmetrical double quantum dot. Eur. Phys. J. Plus **138**, 292 (2023). https://doi.org/10.1140/epjp/s13360-023-03892-8
29. A. Fakkahi, A. Sali, M. Jaouane, R. Arraoui, A. Ed-Dahmouny, Investigation of the nonlinear optical rectification coefficient in a multilayered spherical quantum dot. Opt. Mater. **132**, 112752 (2022). https://doi.org/10.1016/j.optmat.2022.112752
30. M. Jaouane, A. Ed-Dahmouny, A. Fakkahi, R. Arraoui, K. El-Bakkari, H. Azmi, A. Sali, F. Ungan, Investigation of nonlinear optical rectification within multilayer wurtzite InGaN/GaN cylindrical quantum dots under the impact of temperature and pressure. Opt. Mater. **147**, 114711 (2024). https://doi.org/10.1016/j.optmat.2023.114711
31. H. Sayrac, M. Jaouane, A. Ed-Dahmouny, A. Sali, F. Ungan, Modulation of nonlinear optical rectification, second, and third harmonic generation coefficients in n-type quadruple δ-doped GaAs quantum wells under external fields. Phys. B Condens. Matter **690**, 416252 (2024). https://doi.org/10.1016/j.physb.2024.416252
32. *COMSOL Multiphysics*, vol. 5.4 (COMSOL AB, Stockholm, Sweden, 1998). URL www.comsol.com/blogs
33. A. Ed-Dahmouny, H.M. Althib, R. Arraoui, A. Fakkahi, M. Jaouane, H. Azmi, J. El-Hamouchi, K. El-Bakkari, A. Sali, Tuning the optoelectronic properties of GaAs/Al_xGa_{1-x}As core/shell tetrapod quantum dots with a single dopant. Results Phys. **74**, 1–7 (2025). https://doi.org/10.1016/j.rinp.2025.108281

34. A. Fakkahi, R. Arraoui, M. Jaouane, A. Ed-Dahmouny, A. Sali, Central cell correction, pressure and temperature effects on the nonlinear optical rectification and second harmonic generation in a doped GaAs/GaAlAs multi-layer quantum disk. Optik **327**, 172273 (2025). https://doi.org/10.1016/j.ijleo.2025.172273
35. A. Ed-Dahmouny, A. Sali, N. Es-Sbai, R. Arraoui, C.A. Duque, The impact of hydrostatic pressure and temperature on the binding energy, linear, third-order nonlinear, and total optical absorption coefficients and refractive index changes of a hydrogenic donor impurity confined in GaAs/Al_xGa_{1-x}As double quantum dots. Eur. Phys. J. Plus **137**, 784 (2022a). https://doi.org/10.1140/epjp/s13360-022-03002-0
36. A. Ed-Dahmouny, A. Sali, N. Es-Sbai, R. Arraoui, M. Jaouane, A. Fakkahi, K. El-Bakkari, C.A. Duque, Combined effects of hydrostatic pressure and electric field on the donor binding energy, polarizability, and photoionization cross-section in double GaAs/GaAlAs quantum dots. Eur. Phys. J. B **95**, 136 (2022b). https://doi.org/10.1140/epjb/s10051-022-00400-2
37. E. Rosencher, P. Bois, Model system for optical nonlinearities: asymmetric quantum wells. Phys. Rev. B **44**, 11315–11327 (1991). https://doi.org/10.1103/PhysRevB.44.11315
38. A. Ed-Dahmouny, R. Arraoui, M. Jaouane, H.M. Althib, A. Fakkahi, K. El-Bakkari, N. Sali, N. Zeiri, Strain-induced modulations in nonlinear optical rectification of core/shell quantum dots within an MEH-PPV polymer matrix under electric field: a 3D finite-element modeling study. Disc. Appl. Sci. (2024). https://doi.org/10.1007/s42452-024-06257-7
39. O.C. Zienkiewicz, *Isoparametric Elements and Numerical Integration* (Springer Vienna, Vienna, 1972), pp. 30–37. https://doi.org/10.1007/978-3-7091-2973-9_6
40. A. Ed-Dahmouny, N. Zeiri, R. Arraoui, N. Es-sbai, M. Jaouane, A. Fakkahi, A. Sali, The third-order nonlinear optical susceptibility in an ellipsoidal core-shell quantum dot embedded in various dielectric surrounding matrices. Phys. E **153**, 115784 (2023). https://doi.org/10.1016/j.physe.2023.115784

Absorption Coefficient of a Delta-Doped Layer Inside Three Quantum Wells Subjected to an Electric Field

M. Jaouane, A. Ed-Dahmouny, A. Fakkahi, R. Arraoui, K. El-Bakkari, H. Azmi, H. El Ghazi, A. Sali, and C. A. Duque

1 Introduction

The concept of embedding one material within another under precise conditions, like lattice mismatch and energy gap alignment, has opened up a fascinating field in physics. The advancement of growth techniques such as Molecular Beam Epitaxy (MBE), Metal-Organic Chemical Vapor Deposition (MOCVD), and Stranski–Krastanov (SK) growth has enabled the creation of nanoscale structures, commonly known as nanomaterials, including quantum wells (QWs). These structure have shown properties that didn't find in the bulk material, the electronic properties varied dependent on the quantum well dimension. Which have used in different our daily life application such as, solar cell [1, 2], light-emitting transistors [3], light-emitting diodes [4]. The efficiency of nanostructured materials is also related to the materials from which they are made. The III–V family, especially GaAs, has been adopted by various scholars due to its important properties: high electron mobility, direct bandgap, thermal stability, and more [5, 6]. On the other hand, the doping technique plays an important role in semiconductor manufacturing, especially at the nanoscale range. The doping materials involves introducing small amounts of impu-

M. Jaouane · A. Fakkahi · R. Arraoui · K. El-Bakkari · H. Azmi · A. Sali
LPS Laboratory, Faculty of Science Dhar El Mahraz, Sidi Mohamed Ben Abdellah University, Fez, Morocco

A. Ed-Dahmouny (✉)
LaRSI Laboratory, Faculty of Sciences and Technology, Sidi Mohamed Ben Abdellah University, Fez, Morocco
e-mail: ayoub.eddahmouny@usmba.ac.ma

H. El Ghazi
MPIS Group, ENSAM Laboratory, Hassan 2 University, Casablanca, Morocco

C. A. Duque
Grupo de Materia Condensada-UdeA, Instituto de Física, Facultad de Ciencias Exactas y Naturales, Universidad de Antioquia UdeA, Medellín, Colombia

R. En-nadir (ed.), *Quantum Systems at the Nanoscale*, Lecture Notes in Nanoscale Science and Technology 45, https://doi.org/10.1007/978-3-032-06645-9_6

rities (dopants) into a material, to modify its electrical, optical, or structural properties. Delta doping consists of thin layers with high dopant concentration within a spatial limit [7], which can be grown using the MOCVD technique [8, 9]. The delta-doped concept is used to produce a very high carrier concentration. Bass was the first research who use the delta doped layer, produced by MOCVD [10]. Lino et al. have computed the effect of a magnetic field on a δ-doped quantum well [11]. The width of the GaAs layer was varied from 100 to 400 Å. The effect of magnetic field was also used in the GaAs/AlGaAs coupled double quantum wells using the FEniCS Project and Python programming [8]. Jaouane et al. have studied n-type doped layer in GaAs double quantum wells based on the solution of coupling Schrödinger and Poisson equations [12]. Dao et al. have investigated experimentally the photoluminescence in QW made by InGaAs/GaAs with delta layer located either at the barrier region, outside the well and center of center of the QW [13]. The intersuband optical response, absorption coefficient and refractive index of the δ-doped SiGe QW have carried out by Duque et al. [14].

The goal of this paper is to investigate the optical properties of doped, three-coupled quantum wells with an n-type δ-doped layer under an applied electric field along the growth direction. Our calculations use the effective mass approximation to solve the Schrödinger and Poisson equations, implemented using the FEniCS Project. The paper is structured as follows: in Sect. 2, we describe the computational method used to simulate the problem; in Sect. 3, we present and interpret the results; and in the final Sect. 4, we summarize the main findings.

2 Model

The electron quantum systems are derived from the Schrödinger equation in the three quantum wells formed by GaAs materials embedded in $Al_xGa_{1-x}As$, which is typically made by introducing aluminum into GaAs with an x concentration. The system is doped with an N_d concentration in a layer with a δ width located at the z_i position from the center of the system. The Schrödinger equation for an electron in the electric field F parallel to growth direction and the Poisson equation are written as [9],

$$\left(-\frac{\hbar^2}{2}\nabla_z\cdot\left(\frac{1}{m(z)^*}\nabla_z\right)+V_T^{(j)}(z)+eFz\right)\Psi^{(j)}=E^{(j)}\Psi^{(j)} \tag{1}$$

$$\nabla_z\cdot\left(\epsilon\epsilon_0\nabla_z(V_H(z)^{(j)})\right)=-\rho(z)^{(j)} \tag{2}$$

The total potential $V_H(z)$ is related to the confinement potential $V_c(z)+eFz$ and the potential created by the presence of the dopant carriers, expressed as [15]:

$$V_T^{(j)}(z)=95\%\times V_T^{(j-1)}(z)+5\%\times\left(V_c(z)-V_H^{(j-1)}\right) \tag{3}$$

In the first step of the calculation, we assume the total potential equals the confinement potential $V_c(z)$ along side with the potential energy of a charge e in an electric field F. The confinement potential is related to the band gap energy of the GaAs and AlGaAs materials, and is given as follows:

$$V_c(z) = \begin{cases} 0, & z \in GaAs \\ V_0, & z \in AlGaAs \end{cases} \tag{4}$$

The Hartree potential, $V_H(z) = -e\Phi(z)$, represents the electrostatic potential $\Phi(z)$ scaled by the electron charge $-e$.

The charge density $\rho(z)$, which depends approximately only on the electron and donor impurity carriers in a uniformly doped δ-layer with a concentration N_d with an assumption that all the donors are ionized [16]. It is expressed as a function of the z-coordinate, as follows [12]:

$$\rho(z)^{(j)} = \begin{cases} e(N^{(j)}(z) - N_d) & \text{if} \quad z \in \delta \\ e \times N^{(j)}(z) & otherwise. \end{cases} \tag{5}$$

From the eigenvalues and eigenvectors obtained from the Schrödinger Eq. (1), we can calculate the electron charge densities $n(z)$ at the point z as [17],

$$N^{(j)}(z) = \sum_{i=1}^{n_f} \frac{m^* k_b T}{\pi \hbar^2} \log\left[1 + \exp\left(\frac{E_F^{(j)} - E_i^{(j)}}{k_b T}\right)\right] |\Psi_i^{(j)}(z)|^2 \tag{6}$$

From the charge neutrality of our nano-semiconductor system, the number of negative charge carriers must equal the number of positive ones. From this, we can calculate the Fermi energy using this equation,

$$n_d \times \delta = \sum_{i=1}^{n_f} \frac{m^* k_b T}{\pi \hbar^2} \log\left(1 + \exp\left(\frac{E_f^{(j)} - E_i^{(j)}}{k_b T}\right)\right) \tag{7}$$

The calculation iteration, denoted by the index j, continues until the difference between two successive Fermi energy levels is less than 10^{-2} meV. The problem is described by coupled Schrödinger and Poisson equations, which are solved using the FEniCS Project and Python programming within the effective mass approximation. The parameters used in the simulation are listed in Table 1, with $L_d = 6$ nm, $L_b = 4$, $\delta = 2$ nm and $N_d = 4.5 \times 10^{25}$ m^3.

Table 1 List of parameters used in the calculations [15, 18]

Parameter	Value
m_e	9.1093×10^{-31} kg
e	1.6021×10^{-19} C
k_b	1.3806×10^{-23} J K^{-1}
ϵ_0	8.8541×10^{-12} F m
$\hbar$	1.0545×10^{-34} J s
m^*_{GaAs}	$0.067\, m_e$
T	300 K
V_0	0.228 eV
ϵ_r	12.35

3 Results

In this paper, we present the variation of the absorption coefficient and the transition energy of the first five energy levels of a three-quantum-well structure, consisting of a GaAs layer sandwiched between AlGaAs layers, with a δ-layer deposited at position z_i in the middle of the system, under the influence of an external electric field. As presented in Fig. 1.

In Fig. 2, we present transition energy from the ground state and the first five energy levels as a function of the intensity of the external electric field that applied along growth direction with doped layer is located at center of the system $z_i = 0$. It is evident that the transition energy is sensitive to the electric field.

The transition energy E_{1s-2s} decreases with the electric field up to 4.5 kV/cm, indicating that the electric field increases the disparity between the two levels. Beyond this point, E_{1s-2s} starts to fluctuate. The transition energies related to $1s - 3s$ and $1s - 4s$ are less affected by the electric field, while the transitions $1s - 5s$ and $1s - 6s$ are more sensitive to the electric field, decreasing as F increases, in this situation the electron is less stable. These behaviors can be explained physically as

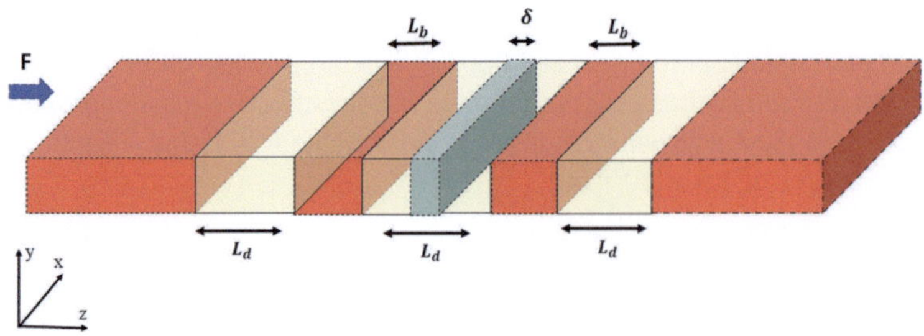

Fig. 1 An illustration of three coupled quantum wells with a doped layer of width δ under an applied electric field F

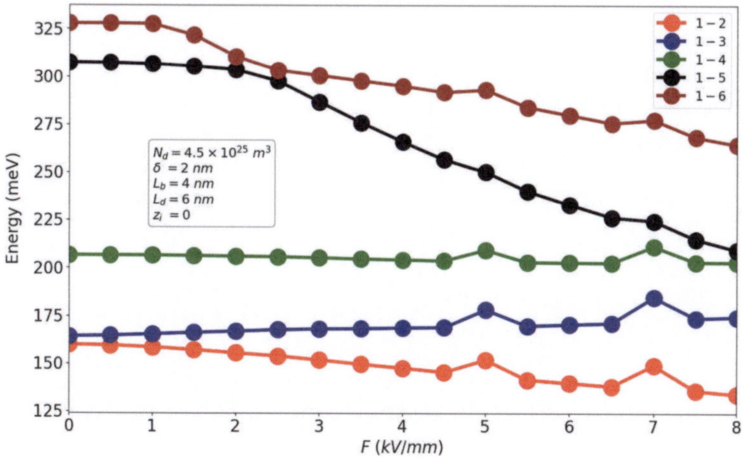

Fig. 2 The transition energy between the ground state and the first four excited states as a function of the electric field

follows: the electric field deforms the confining potential, allowing the electron to exit the well region with less energy, which changes the electron density. Additionally, the electric field displaces the electron's wavefunction. It also alters the electron-impurity interaction by increasing the separation between opposite charges. These results can be found in the case of electron with one impurity [19].

The absorption coefficient associated with the transition from the ground state to the first five excited states as a function of incident photon energy is presented in Fig. 3, where the doped region is centered at $z_i = 0$, in the absence of electric field $F = 0$ kV/cm. The absorption coefficient increases with $\hbar\omega$ until it reaches a maximum at the resonant energy, corresponding to the transition energy, after which it decreases to lower values. Furthermore, the absorption coefficient exhibits a blueshift as the final states change from 2 to 6. With an external electric field applied along the growth direction, as shown in Fig. 1, we investigated the variation of the absorption coefficient of first transition $1-2$ with respect to the photon energy, $\hbar\omega$, for two different positions of the δ layer: $z_i = 0$ and $z_i = \frac{L_d+L_b}{2}$.

For the case of $z_i = 0$, the absorption coefficient shows a field-dependent variation, with a redshift as the electric field increases. The peak position is related to the transition energy, a behavior predicted by Fig. 2. After the electric field reaches 3 kV/cm, the transition becomes less influenced by the field. Additionally, the amplitude of the peak position is related to the wavefunction distribution across the system. When the electric field is increased within a range less than $F = 5$ kV/cm, the amplitude rises with F; beyond this range, however, the amplitude begins to diminish (Fig. 4).

In the case of Fig. 5, the right quantum well is doped with an n-type δ-layer located at $z_i = \frac{L_d+L_b}{2}$. As we can see from this figure, the absorption coefficient is significantly different from the case of $z_i = 0$, being approximately one order of

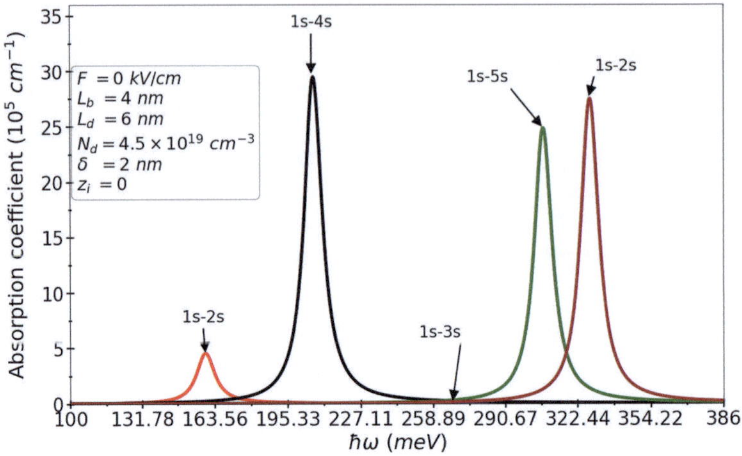

Fig. 3 The variation of the absorption coefficient as a function of the incident photon energy associated with the different transitions, with the delta layer located at $z_i = 0$

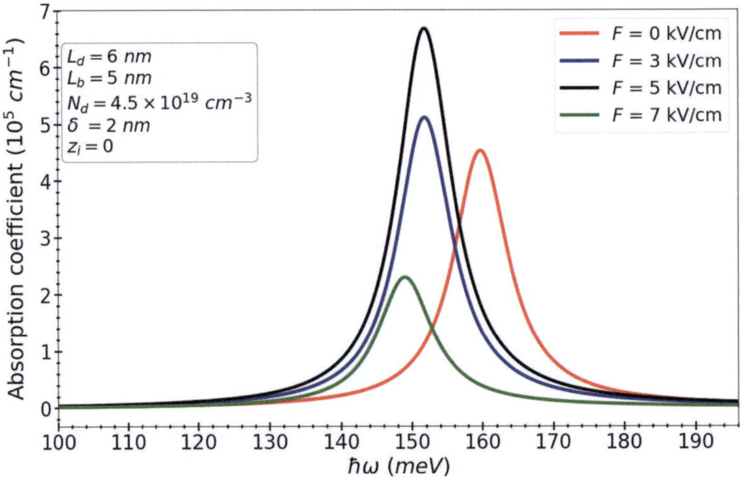

Fig. 4 The variation of the absorption coefficient as a function of the incident photon energy for different values of the electric field, with the delta layer located at $z_i = 0$

magnitude lower. As the electric field increases, it slightly shifts the peak position to smaller photon energies, this redshift is due to the decrease of the differences between the energy levels with the electric field. The absorption coefficient amplitude decreases, eventually disappearing for field values larger than 5 kV/cm. These shows the same as in the work done by Karabulut et al. for the case of one impurity in the spherical quantum dot [20].

In the following, we show how the absorption coefficient is affected by the delta-doped position in the absence and presence of electric field values. In the absence

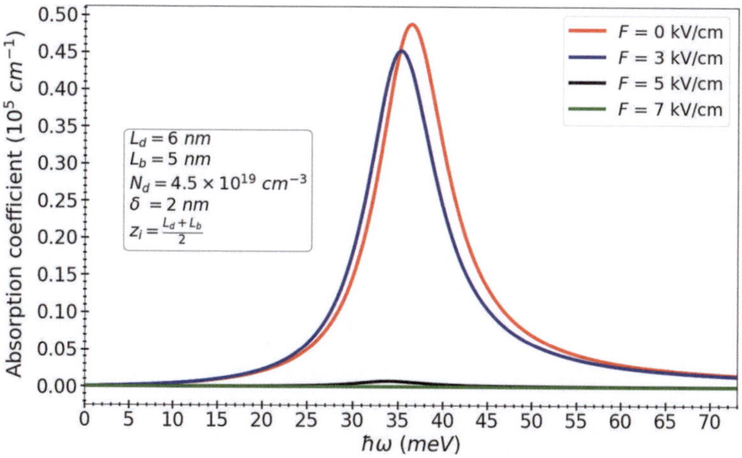

Fig. 5 The variation of the absorption coefficient as a function of the incident photon energy for different values of the electric field, with the delta layer located at $z_i = \frac{L_d+L_b}{2}$

Table 2 The transition energy as a function of the electric field and impurity position

Transition energy	$F = 0$ kV/cm (meV)	$F = 50$ kV/cm (meV)
$E_{1s,2s}(z_i = 0)$	158.60	149.6
$E_{1s,2s}(z_i = L_d + L_b)$	178.7	172.9
$E_{1s,2s}(z_i = -L_d - L_b)$	178.7	179
$E_{1s,2s}(z_i = (L_d + L_b)/2)$	35.5	34.2
$E_{1s,2s}(z_i = -(L_d + L_b)/2)$	35.5	32.6

of the electric field, as the delta layer changes from the center of the quantum well (QW) to the right or left QW, it means that the position where the donor impurities are located around z_i leads to a blueshift of the absorption coefficient. This shift falls within the mid-infrared (MIR) to near-infrared (NIR) region and becomes identical for two different positions, $z_i = L_d + L_b$ and $z_i = -L_d - L_b$, due to the symmetry of the system. However, this result no longer holds in the presence of an electric field. The absorption coefficient for $z_i = 0$ and $z_i = L_d + L_b$ is blueshifted by the electric field, while it remains approximately unchanged for $z_i = -L_d - L_b$.

In Fig. 7, we plot the absorption coefficient versus photon energy for two different positions in the absence and presence of an electric field. In the absence of an electric field, the absorption coefficient is identical for $z_i = \frac{L_d+L_b}{2}$ and $z_i = -\frac{L_d+L_b}{2}$. The peaks occur at $E_{1s,2s} = 35.5$ meV, which corresponds to $\lambda = 35.45$ μm. The presence of an electric field significantly affects the absorption coefficient amplitude and causes a redshift. These opposite effects are shown for other positions in Fig. 6 (Table 2).

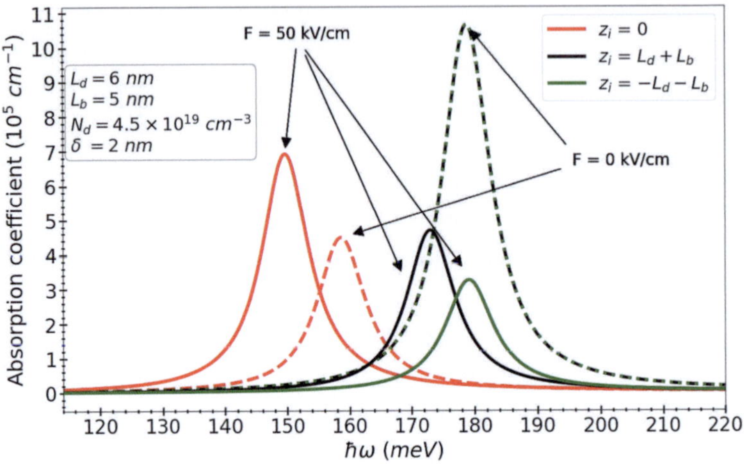

Fig. 6 The variation of the absorption coefficient as a function of the incident photon energy, associated with different delta layer positions, right QW, center QW, and left QW, at two different values of $F = 0$ kV/cm and $F = 50$ kV/cm

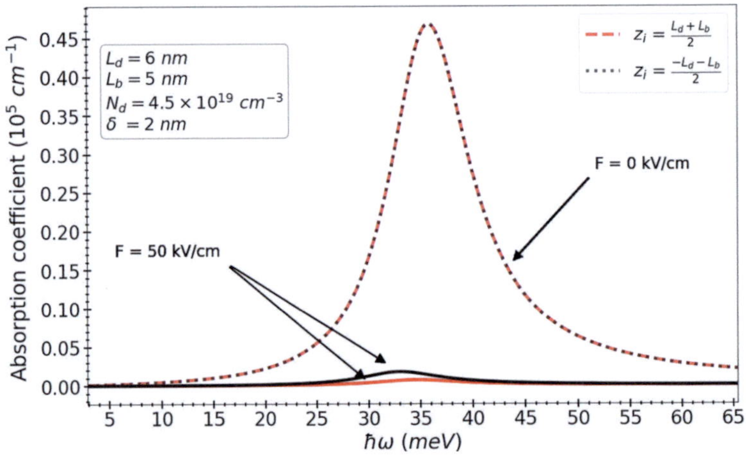

Fig. 7 The variation of the absorption coefficient as a function of the incident photon energy, associated with different delta layer positions (right center barrier and left center barrier), at two different values of $F = 0$ kV/cm and $F = 50$ kV/cm

4 Conclusion

In conclusion, the study presented in this chapter successfully demonstrates the solution of the coupled Schrödinger-Poisson equation for a system of three coupled quantum wells with an n-type δ-layer under the influence of an external electric field. By employing the FEniCS Project within the effective mass approximation, we observed that the transition energy of the first five states is significantly dependent on the electric field. This, in turn, affects both the position and amplitude of the absorption coefficient, which are influenced by the location of the δ-layer. These findings offer valuable insights into the manipulation of electronic and optical properties, which are essential for optimizing the performance of devices such as transistors and solar cells. Future work will explore further the implications of these results in the design and improvement of quantum devices and materials.

Acknowledgements CAD is grateful to the Colombian Agencies: CODI-Universidad de Antioquia (Estrategia de Sostenibilidad de la Universidad de Antioquia and projects "Complejos excitónicos y propiedades de transporte en sistemas nanométricos de semiconductores con simetría axial basados en GaAs e InAs", and "Nanoestructuras semiconductoras con simetría axial basadas en InAs y GaAs para aplicaciones en electrónica ultra e híper rápida") and Facultad de Ciencias Exactas y Naturales-Universidad de Antioquia (CAD exclusive dedication project 2025–2026).

Data Availability The corresponding author can provide all files related to tables, figures, and code on request.

Declaration of Competing Interest The authors have no conflicts of interest to disclose.

References

1. M. Jaouane, A. Fakkahi, A. Ed-Dahmouny, K. El-Bakkari, A.T. Tuzemen, R. Arraoui, A. Sali, F. Ungan, Modeling and simulation of the influence of quantum dots density on solar cell properties. Eur. Phys. J. Plus **138** (2023). https://doi.org/10.1140/epjp/s13360-023-03736-5
2. Z. Arefinia, D.P. Samajdar, Fundamental limiting efficiency and intrinsic loss components of quantum-wire intermediate-band solar cells. Phys. Rev. Appl. **14** (2020). https://doi.org/10.1103/physrevapplied.14.064061
3. M. Kumar, S.-Y. Ho, S.-J. Hsu, P.-C. Li, S.-W. Chang, C.-H. Wu, Current gain enhancement at high-temperature operation of triple-quantum-well heterojunction bipolar light-emitting transistor for smart thermal sensor application. IEEE Trans. Electron Devices **71**, 896–903 (2024). https://doi.org/10.1109/ted.2023.3339084
4. B. Zeinalvand Farzin, S.B. Seyedein Ardebili, T.I. Kang, J.S. Kim, P.D. Nguyen, S.J. Lee, Photoreflectance analysis of InAsPSb/InGaAs multi-quantum well led structures with different well/barrier numbers. Photonics **11**, 277 (2024). https://doi.org/10.3390/photonics11030277
5. A. Ed-Dahmouny, A. Sali, N. Es-Sbai, R. Arraoui, M. Jaouane, A. Fakkahi, K. El-Bakkari, C.A. Duque, Combined effects of hydrostatic pressure and electric field on the donor binding energy, polarizability, and photoionization cross-section in double GaAs/GaAlAs quantum dots. Eur. Phys. J. B **95**, 136 (2022a). https://doi.org/10.1140/epjb/s10051-022-00400-2
6. A. Ed-Dahmouny, A. Sali, N. Es-Sbai, R. Arraoui, C.A. Duque, The impact of hydrostatic pressure and temperature on the binding energy, linear, third-order nonlinear, and total optical absorption coefficients and refractive index changes of a hydrogenic donor impurity confined in GaAs/$Al_xGa_{1-x}As$ double quantum dots. Eur. Phys. J. Plus **137**, 784 (2022b). https://doi.org/10.1140/epjp/s13360-022-03002-0
7. H. Sayrac, M. Jaouane, A. Ed-Dahmouny, A. Sali, F. Ungan, Modulation of nonlinear optical rectification, second, and third harmonic generation coefficients in n-type quadruple δ-doped GaAs quantum wells under external fields. Phys. B Condens. Matter **690**, 416252 (2024). https://doi.org/10.1016/j.physb.2024.416252

8. M. Jaouane, A. Ed-Dahmouny, A. Fakkahi, R. Arraoui, H. Azmi, A. Sali, M. El Sayed, A. Samir, Quantum landau levels in n-type modulation-doped GaAs/AlGaAs coupled double quantum wells. J. Magn. Magn. Mater. **608**, 172406 (2024a). https://doi.org/10.1016/j.jmmm.2024.172406
9. M. Jaouane, A. Ed-Dahmouny, R. Arraoui, H.M. Althib, A. Fakkahi, H. El Ghazi, A. Sali, M.N. Murshed, N. Zeiri, Delta-doping modulation of three quantum wells under the influence of an electric field. APL Mater. **12** (2024b). https://doi.org/10.1063/5.0231797
10. S. Bass, Silicon and germanium doping of epitaxial gallium arsenide grown by the trimethylgallium-arsine method. J. Cryst. Growth **47**, 613–618 (1979). https://doi.org/10.1016/0022-0248(79)90002-2
11. A.T. Lino, E.K. Takahashi, L.M.R. Scolfaro, J.R. Leite, Behavior of carriers in δ-doped quantum wells under in-plane magnetic fields. Int. J. Quant. Chem. **60**, 1559–1566 (1996). https://doi.org/10.1002/(sici)1097-461x(1996)60:7<1559::aid-qua37>3.0.co;2-w
12. M. Jaouane, A. Ed-Dahmouny, A. Fakkahi, R. Arraoui, K. El-Bakkari, H. Azmi, A. Sali, C. Duque, n-type doping modulation of double GaAs/AlGaAs quantum wells. Comput. Mater. Sci. **238**, 112930 (2024). https://doi.org/10.1016/j.commatsci.2024.112930
13. L.V. Dao, M. Gal, G. Li, C. Jagadish, Photoluminescence in delta-doped InGaAs/GaAs single quantum wells. J. Appl. Phys. **87**, 3896–3899 (2000). https://doi.org/10.1063/1.372431
14. C. Duque, V. Akimov, R. Demediuk, V. Belykh, A. Tiutiunnyk, A. Morales, R. Restrepo, M. Mora-Ramos, O. Fomina, V. Tulupenko, Intersubband linear and nonlinear optical response of the delta-doped SiGe quantum well. Superlattices Microstruct. **87**, 125–130 (2015). https://doi.org/10.1016/j.spmi.2015.05.039
15. M. Jaouane, A. Ed-Dahmouny, A. Fakkahi, R. Arraoui, H. Azmi, A. Sali, M. El Sayed, A. Samir, Quantum landau levels in n-type modulation-doped GaAs/AlGaAs coupled double quantum wells. J. Magn. Magn. Mater. **608**, 172406 (2024). https://doi.org/10.1016/j.jmmm.2024.172406
16. T. Sahu, J. Patnaik, Electron transport mobility in a δ-doped double quantum well structure. J. Appl. Phys. **88**, 2658–2664 (2000). https://doi.org/10.1063/1.1286727
17. L.R. Ram-Mohan, *Finite Element and Boundary Element Applications in Quantum Mechanics* (Oxford University Press, Oxford)
18. A. Ed-Dahmouny, H.M. Althib, R. Arraoui, A. Fakkahi, M. Jaouane, H. Azmi, J. El-Hamouchi, K. El-Bakkari, A. Sali, Tuning the optoelectronic properties of GaAs/Al_xGa_{1-x}As core/shell tetrapod quantum dots with a single dopant. Results Phys. **74**, 1–7 (2025). https://doi.org/10.1016/j.rinp.2025.108281
19. M. Jaouane, A. Sali, A. Ezzarfi, A. Fakkahi, R. Arraoui, Study of hydrostatic pressure, electric and magnetic fields effects on the donor binding energy in multilayer cylindrical quantum dots. Phys. E Low-Dimens. Syst. Nanostruct. **127**, 114543 (2021). https://doi.org/10.1016/j.physe.2020.114543
20. I. Karabulut, S. Baskoutas, Linear and nonlinear optical absorption coefficients and refractive index changes in spherical quantum dots: effects of impurities, electric field, size, and optical intensity. J. Appl. Phys. **103** (2008). https://doi.org/10.1063/1.2904860

Stark-Shift, Binding Energy, Diamagnetic Response, Polarizability and Dipole Moment of Shallow Donor-Impurity in Core/Shell Quantum Disk

Ibrahim Maouhoubi

1 Introduction

In the realm of condensed matter physics and materials science, the exploration of low-dimensional systems has yielded a plethora of captivating phenomena and promising technological applications [1, 2]. These systems, characterized by reduced dimensions in one or more directions, offer a unique platform to investigate and harness novel physical and chemical properties [3–8]. Among these intriguing systems, the focus on semiconductor materials has been particularly pronounced, with their potential to revolutionize various optoelectronic and electronic devices [9–12].

The subject of this study delves into the intricate world of low-dimensional semiconductors, specifically directed towards the investigation of a shallow donor-impurity embedded within a GaAs/$Al_{0.3}Ga_{0.7}As$ core/shell quantum disk (QDisk) under the influence of a lateral applied electric field. This multifaceted exploration encompasses a comprehensive examination of the Stark-shift, binding energy, diamagnetic response, polarizability, and dipole moment of the donor-impurity system. The incorporation of a lateral directed applied electric field introduces an additional layer of complexity, enabling the manipulation of the system's behavior and response. The GaAs/$Al_{0.3}Ga_{0.7}As$ core/shell quantum disk architecture serves as a compelling case study due to its rich potential for tailored optical and electrical properties [4, 13–15]. The core/shell structure, consisting of a GaAs core enveloped by an $Al_{0.3}Ga_{0.7}As$ shell, exhibits intriguing features that can be harnessed for a range of applications spanning from optoelectronic devices, such as solar cells [2, 16] and lasers [17], to other emerging fields like biosensing and nanocomposites [18–21].

I. Maouhoubi (✉)
LPS, Department of Physics, Sidi Mohamed Ben Abdellah University, Fez, Morocco
e-mail: ibrahim.maouhoubi@usmba.ac.ma

R. En-nadir (ed.), *Quantum Systems at the Nanoscale*, Lecture Notes in Nanoscale Science and Technology 45, https://doi.org/10.1007/978-3-032-06645-9_7

The investigation of the aforementioned properties within this core/shell quantum disk framework holds profound implications. A detailed analysis of the Stark-shift provides insights into how the system responds to external electric fields [22], while the exploration of the binding energy [23] sheds light on the stability of the donor-impurity interaction. Additionally, the diamagnetic response, polarizability [1, 24], and dipole moment elucidate the magnetic and electric properties of the system under lateral applied electric fields.

By undertaking this study, we aim to contribute to the evolving understanding of low-dimensional semiconductor systems and their intricate behaviors under the influence of external fields. The insights gained from this investigation hold the potential to enrich our comprehension of these systems and offer new avenues for designing and engineering advanced optoelectronic and electronic devices.

2 Theoretical Framework

In this chapter, our focus is on investigating the influence of a $\vec{F}_{(x,y)}$-field and its orientation on the electronic and physical properties of a negative charge confined within an inhomogeneous quantum disk in the existence of a donor defect. We begin by deriving the equation of the Hamiltonian, which translates the negative charge's behavior under the influence of the electric field applied laterally. Subsequently, we delve into the methods utilized to solve the Schrödinger equation, which allows us to determine the eigenvalues and eigenvectors associated with the system. These eigenvalues and eigenvectors provide crucial insights into the energy levels and wavefunctions of the confined negative charge in the existence of the defect and the lateral electric field.

2.1 The Hamiltonian of System

In our study, we are investigating a two-dimensional (2D) circular core/shell quantum dot (QD) made out of GaAs/GaAlAs with inner and outer radii represented by R_{in} and R_{out}, respectively. To understand the behavior of the system under examination, we apply an external lateral electric field within the (xOy) plane, and its proper angle is denoted by φ_F. The electron's position is described by polar coordinates (ρ, φ), while the defect's position is specified by (ρ_D, ρ_{De}), and ρ_{De} represents the electron–defect separation. Notably, in this research, we assume that ρ_D is oriented along the x-direction and is equal to $\rho_D = (R_{in} + R_{out})/2$. This configuration helps us to examine the influence of the $\vec{F}_{(x,y)}$-field and its direction on the electronic and physical properties of the investigated system, all these indication are depicted in see Fig. 1.

The Hamiltonian in this system accounts for the correlation between the electron and impurity. It can be written as the sum of four components,

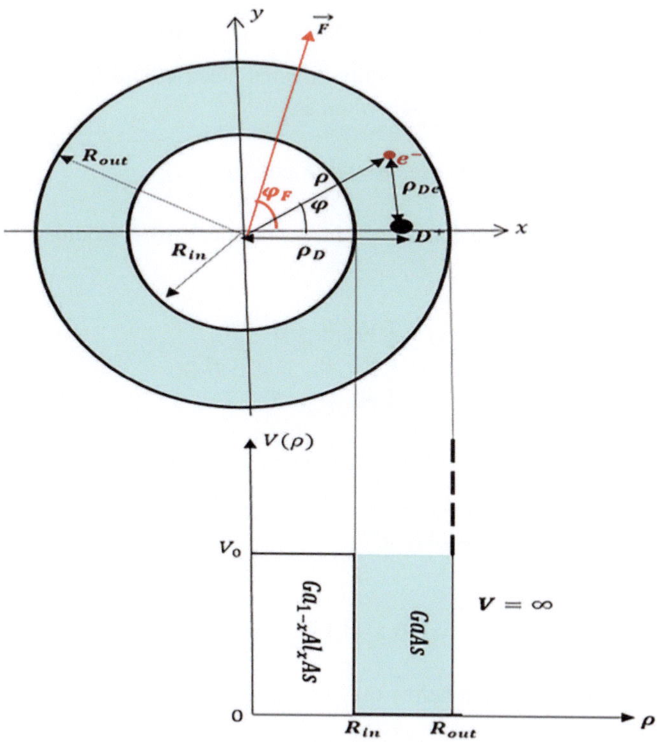

Fig. 1 Schematic design of 2D thin quantum dots core/shell with adjustable $\vec{F}_{(x,y)}$-field

$$H = -\frac{\hbar^2}{2m^*_{w,b}}\left(\frac{1}{\rho}\frac{\partial}{\partial\rho}\left(\rho\frac{\partial}{\partial\rho}\right) + \frac{1}{\rho^2}\frac{\partial^2}{\partial\varphi^2}\right) + V_{De} + W + V(\rho) \tag{1}$$

The symbol $m^*_{w,b}$ represents the effective mass of electron for the core (well) and shell (barrier) materials, which in this case are GaAs and $Al_xGa_{1-x}As$, respectively. The parameter 'x' represents the aluminum content in the $Al_xGa_{1-x}As$ material. The electron's effective mass in the well and barrier materials is influenced by the concentration of aluminum, and this dependence can be described by a specific function or relationship [1].

$$m^*_{w,b} = \begin{cases} m^*_w(GaAs) = 0.067m_0 \\ m^*_b(Ga_{1-x}Al_xAs) = (0.067 + 0.835x)m_0 \end{cases} \tag{2}$$

The electrostatic repulsion happened between defect-electron is described by the Coulomb potential, denoted as V_{De}, which is given by $V_{De} = \frac{e^2}{\varepsilon_0\left|\vec{\rho}-\overrightarrow{\rho_D}\right|}$. Here, ε_0 represents the static dielectric constant, e is the electronic charge, and $\left|\vec{\rho}-\overrightarrow{\rho_D}\right|$ is the distance between the defect and the electron, given as $\left|\vec{\rho}-\overrightarrow{\rho_D}\right| =$

$\sqrt{\rho^2 + \rho_D^2 - 2\rho\rho_D \cos\varphi}$. Additionally, the electric dipole energy operator, represented by W, is introduced for a laterally directed electric field $\vec{F}$. In this situation, the expression of the operator W is written as.

$$W = e\vec{F} \cdot \vec{\rho} = e\rho F \cos(\varphi_F - \varphi) \tag{3}$$

The confinement potential barrier height, denoted as V(ρ), is modulated according to the following expression:

$$V(\rho) = \begin{cases} V_0, \ 0 \le \rho \le R_{in} \\ 0, \ \ R_{in} \le \rho \le R_{out} \\ \infty, \ \rho > R_{out} \end{cases} \tag{4}$$

The parameter ΔE_g depicts the fluctuations in energy band gaps between the core and shell materials, where $\Delta E_g = E_g^{Ga_{1-x}Al_xAs} - E_g^{GaAs}$. The parameter V_0 is then defined as $V_0 = 0.6\Delta E_g(x)$, and It is mathematically represented in analytical form as follows [1]:

$$V_0 = 0.6 \times \left(1.04x + 0.47x^2\right) \tag{5}$$

The Schrödinger equation below may be solved to get the defect's energy in its lowest state:

$$\mathcal{H}_0(\rho, \varphi, F, x) \cdot \psi_0(\rho, \varphi, F, x) = E_0\psi_0(\rho, \varphi, F, x) \tag{6}$$

Taking into account both the contributions from the $\vec{F}_{(x,y)}$-field and the defect-electron Coulomb contact, Eq. (6) can be obtained as:

$$\mathcal{H}(\rho, \varphi, F, \rho_i, x) \cdot \psi_i(\rho, \varphi, F, \rho_i, x) = E_t\psi_i(\rho, \varphi, F, \rho_i, x) \tag{7}$$

2.2 *Finite Difference Method*

As mentioned earlier, we have employed the Finite Difference Method (FDM) to numerically solve the Schrödinger equation for the system. To discretize the 2D space, a lattice mesh-grid with dimensions $\left(N_\rho \times N_\varphi\right)$ is used. The intervals $[R_{in}; R_{out}]$ and [0; 2π] have been subdivided into $\left(N_\rho + 1\right)$ and $\left(N_\varphi + 1\right)$ segments, respectively. In this section, we will utilize the finite difference method to find numerical solutions to the Schrödinger equation.

The FDM was utilized to numerically explain the Schrödinger calculation, and in doing so, we applied specific boundary conditions. A 2D-mesh with $\left(N_\rho, N_\varphi\right)$

nodes was considered for the calculations. Each point (i, j) on this uniform mesh corresponds to a position along the quantum disk, where $1 \leq i \leq N_\rho$ and $1 \leq j \leq N_\varphi$. To represent the radial coordinate ρ and azimuthal angle φ, we set $\rho = i\Delta\rho$ and $\varphi = j\Delta\varphi$, respectively. Here, The grid points are labeled by integers i and j, while the spacing between them is denoted as $\Delta\rho = \frac{R_{out}-R_{in}}{N_\rho+1}$ and $\Delta\varphi = \frac{2\pi}{N_\varphi+1}$. These discretizations help in performing the numerical calculations and solving the Schrödinger equation efficiently.

Δρ corresponds to the distance separating consecutive points on the grid along the ρ direction. Similarly, Δφ represents the angle between successive points on the grid in the rotational φ direction. At a specific grid point (i, j), the discretization of the system gives rise to the following recurrence relation.

$$\begin{aligned}
&\frac{-\hbar^2}{2m^*_{w,b}}\Bigg[\left(\frac{1}{(\Delta\rho)^2}-\frac{1}{2i(\Delta\rho)^2}\right)\psi_{i-1,j}\\
&+\left(\frac{-2}{(\Delta\rho)^2}-\frac{2}{i^2(\Delta\rho)^2(\Delta\varphi)^2}-\frac{2m^*_{w,b}}{\hbar^2}\left(V^{i,j}_{De}+W^{i,j}+V(\rho)^{i,j}\right)\right)\psi_{i,j}\\
&+\left(\frac{1}{(\Delta\rho)^2}+\frac{1}{2i(\Delta\rho)^2}\right)\psi_{i+1,j}+\frac{1}{i^2(\Delta\rho)^2(\Delta\varphi)^2}\psi_{i,j-1}\\
&+\frac{1}{i^2(\Delta\rho)^2(\Delta\varphi)^2}\psi_{i,j+1}\Bigg]=E\psi_{i,j}
\end{aligned} \tag{8}$$

In order to calculate the eigenvectors and eigenvalues, we apply specific boundary conditions at the superficial and midpoint of the system, following Dirichlet-type circumstances, while adopting Neumann-type circumstances along the uprising axis. This results in equations for $\psi_{N_\rho+1,j} = \psi_{i,N_\varphi+1} = \psi_{0,j} = \psi_{i,0}$. The energy equation can be expressed more simply in this context.

$$\delta_{ij}\psi_{i,j} + \gamma_{ij}\psi_{i+1,j} + u_{ij}\psi_{i-1,j} + \alpha_{ij}\psi_{i,j-1} + \beta_{ij}\psi_{i,j+1} = E\psi_{i,j}, \tag{9}$$

where

$$\delta_{ij} = -\frac{\hbar^2}{2m^*_{w,b}}\left(\frac{2}{(\Delta\rho)^2}+\frac{2}{i^2(\Delta\rho)^2(\Delta\varphi)^2}\right)+V^{i,j}_{De}+W^{i,j}+V(\rho)^{i,j} \tag{10}$$

$$\gamma_{ij} = \frac{-\hbar^2}{2m^*_{w,b}}\left(\frac{1}{(\Delta\rho)^2}+\frac{1}{2i(\Delta\rho)^2}\right) \tag{11}$$

$$u_{ij} = \frac{-\hbar^2}{2m^*_{w,b}}\left(\frac{1}{(\Delta\rho)^2}-\frac{1}{2i(\Delta\rho)^2}\right) \tag{12}$$

$$\alpha_{ij} = \beta_{ij} = \frac{-\hbar^2}{2m^*_{w,b}}\left(\frac{1}{i^2(\Delta\rho)^2(\Delta\varphi)^2}\right) \tag{13}$$

So, Eq. (9) can be converted into a matrix representation as follows:

$$\begin{pmatrix} \delta_{11} & \gamma_{11} & 0 & \cdots & \alpha_{11} & 0 & 0 \\ u_{11} & \delta_{12} & \gamma_{12} & 0 & \cdots & \alpha_{12} & 0 \\ 0 & u_{12} & \ddots & \ddots & \ddots & \vdots & \vdots \\ \vdots & 0 & \ddots & \delta_{ij} & \ddots & 0 & \alpha_{N_\rho N_\varphi} \\ \beta_{11} & \vdots & \ddots & \ddots & \ddots & \gamma_{N_\rho-2,N_\varphi} & 0 \\ 0 & \beta_{12} & \cdots & 0 & u_{N_\rho-1,N_\varphi} & \delta_{N_\rho-1,N_\varphi} & \gamma_{N_\rho-1,N_\varphi} \\ 0 & 0 & \cdots & \beta_{N_\rho N_\varphi} & 0 & u_{N_\rho,N_\varphi} & \delta_{N_\rho,N_\varphi} \end{pmatrix} \begin{pmatrix} \psi^{n,l}_{1,1} \\ \psi^{n,l}_{1,2} \\ \vdots \\ \psi^{n,l}_{i,j} \\ \vdots \\ \psi^{n,l}_{N_\rho-1,N_\varphi} \\ \psi^{n,l}_{N_\rho,N_\varphi} \end{pmatrix}$$

$$= E^{n,l} \begin{pmatrix} \psi^{n,l}_{1,1} \\ \psi^{n,l}_{1,2} \\ \vdots \\ \psi^{n,l}_{i,j} \\ \vdots \\ \psi^{n,l}_{N_\rho-1,N_\varphi} \\ \psi^{n,l}_{N_\rho,N_\varphi} \end{pmatrix} \tag{14}$$

To calculate the energy, the Hamiltonian matrix is precisely resolved using the FDM through the following algorithm:

- Diagonalize the matrix H to determine and order the electron energies $E_e^{n,l}\left(1 < n < N_\rho \cdot N_\varphi \; and \; l = 2k + 1\right)$.
- Introduce each energy $E_e^{n,l}$ into the equation $\mathcal{H}(\rho, \varphi, F, \rho_i, \xi) \cdot \psi^{n,l}_{i,j}(\rho, \varphi, F, \rho_i, \xi) = E_t \psi^{n,l}_{i,j}(\rho, \varphi, F, \rho_i, \xi)$.
- Solve the equation to obtain the associated wave functions $\psi^{n,l}_{i,j}$ at all points of the mesh.

2.3 Electronics and Physical Properties

Dipole Moment is a quantum mechanical property indicating charge distribution separation within a system. While the polarizability is quantum property describing how easily a system's electron cloud adjusts to an external electric field. The expression describing the connection among the electric dipole (μ) and the polarizability (α_p) is given as below [23]:

$$\alpha_p = \frac{\mu}{F} \tag{15}$$

The electric field strengths are represented by F, and the dipole moment (μ) can be calculated using the following expression:

$$\mu = -e\left[\left\langle\psi(\rho,\varphi)\middle|\vec{F}\cdot\vec{\rho}\middle|\psi(\rho,\varphi)\right\rangle_{F\neq 0} - \left\langle\psi(\rho,\varphi)\middle|\vec{F}\cdot\vec{\rho}\middle|\psi(\rho,\varphi)\right\rangle_{F=0}\right] \tag{16}$$

Furthermore, the expression for the dipole moment (μ) can be prolonged by incorporating the system's coordinates, resulting in the ensuing formulation:

$$\mu = -e\left[\left\langle\vec{F}\cdot\vec{\rho}\right\rangle^{F\neq 0}_{\psi(\rho,\varphi)} - \left\langle\vec{F}\cdot\vec{\rho}\right\rangle^{F=0}_{\psi(\rho,\varphi)}\right] \tag{17}$$

The expression for the magnetization response of a donor impurity restricted inside our investigated nanostructure is provided as follows [25]:

$$\chi_{dia} = -\frac{e^2}{6m^*_{w,b}\varepsilon_0 c^2}\left\langle(\vec{\rho}-\vec{\rho}_D)^2\right\rangle \tag{18}$$

$\left\langle(\vec{\rho}-\vec{\rho}_D)^2\right\rangle$ represents the mean square of the defect-electron separation.

$$\left\langle(\vec{\rho}-\vec{\rho}_D)^2\right\rangle = \left\langle\rho^2+\rho_D^2-\rho\rho_D\cos(\varphi_F-\varphi)\right\rangle \tag{19}$$

The ground-level ionization energy (E_b) of the donor is determined by calculating the energy change among the system without the impurity and the system with the impurity, where the energy has been minimized. This binding energy can be mathematically represented as follows:

$$E_b = E_0 - E_t \tag{20}$$

Furthermore, the Stark Change (ΔE_{ss}) refers to the change in energy levels caused by an applied external electric field. This shift occurs due to the interaction between the electric field and the charged particles within the quantum dot, leading to a displacement of the energy levels. The Stark shift can have significant implications for the behavior and properties of the confined particles, influencing aspects such as binding energy, excitations, and overall system response. This shift in energy can be mathematically expressed as follows:

$$\Delta E_{ss} = E_t(F\neq 0) - E_t(F=0) \tag{21}$$

3 Results and Discussion

Throughout this part, we examine a shallow defect located within a core/shell quantum system's numerous features, namely the stark-shift, binding energy, diamagnetic response, polarizability, and dipole moment. The defect is located at a specific x-site, given by $(R_{out} - R_{in})/2$, where $\varphi_i = 0°$ (refer to Fig. 1). Here, R_{out} designates the Shell's diameter and R_{in} represents the Core's diameter of the quantum disk. It is important to note that we always maintain the condition $R_{in} \leq R_{out}$. The Shell is composed of a fixed Aluminum concentration ($Ga_xAl_{1-x}As$) set to x = 30%, corresponding to $Ga_{0.7}Al_{0.3}As$.

To investigate the quantum confinement effect resulting from the dimension reduction and geometry of the semiconductor QD on the confined electron within the nanostructure, we examined the probability density of the electron. In Fig. 2, we presented a series of contours in a selected plane corresponding to an $\vec{F}_{(x,y)}$-field oriented alongside the x-axis. From this figure, it is evident that for larger shell radii ($R_{out} = 20$ nm), the impact of the $\vec{F}_{(x,y)}$-field becomes very important. Consequently, the negative charge is restricted towards the opposite direction of the $\vec{F}_{(x,y)}$ field thanks to the stark change. Conversely, for smaller R_{out} radii ($R_{out} = 10$ nm), the geometric confinement dominates, outweighing the contribution of the $\vec{F}_{(x,y)}$-field.

3.1 The Binding Energy

Prior to discussing the outcomes, it is significant to note that the $\frac{R_{in}}{R_{out}}$ fluctuations within the interval of [0, 1]. Consequently, two distinct cases arise: When the ratio is close to zero, denoted as ($R_{in} \ll R_{out}$), the system resembles a regular QD, where ΔR is significantly superior associated to R_{in}. The outcomes gotten in this situation concerning the ionization energy align well with the findings of Khamkhami and his teamwork [26]. The additional situation occurs when $\frac{R_{in}}{R_{out}} \sim 1$, meaning ($R_{in} \approx R_{out}$). This relates to an extremely narrow circular QWW, and it is observed that the binding energy have a tendency to a minimum value of $E_b = 0.015$ eV. Figure 3 illustrates the correlation energy concerning the $\frac{R_{in}}{R_{out}}$ (in effective form) for different selected orientations of the transverse $\vec{F}_{(x,y)}$-field φ_F. The plot clearly demonstrates that is notably impacted by the system's size, the strength of the $\vec{F}_{(x,y)}$-field, and its specific angle relative to the positive x-direction projection.

In general, irrespective of the electric field's intensity or orientation, a consistent increase in binding energy is noted as the ratio of radii approaches a critical threshold around 0.06 eV for $\frac{R_{in}}{R_{out}} \sim 0.65$. Beyond this point, it decreases quickly and reaches a steady region for $\frac{R_{in}}{R_{out}} > 0.9$. This phenomenon is elucidated by the fact that with an increase in the $\frac{R_{in}}{R_{out}}$ of radii (specifically, as R_{in} increases), the distance between the defect and electron diminishes, resulting in an enhancement of the ground state binding energy. However, The decline in binding energy is linked to the decrease in quantum confinement caused by electron tunneling moving from the Core (GaAs) into the Shell region ($Ga_xAl_{1-x}As$), owing to the electron's penetration through the finite potential barrier.

Moreover, it has been observed that as the φ_F changes from $\varphi_F = 0°$ to $\varphi_F = 180°$, The binding energy is heightened, a consequence attributed to the diminished electrostatic repulsion stemming from a reduced separation between the defect and the

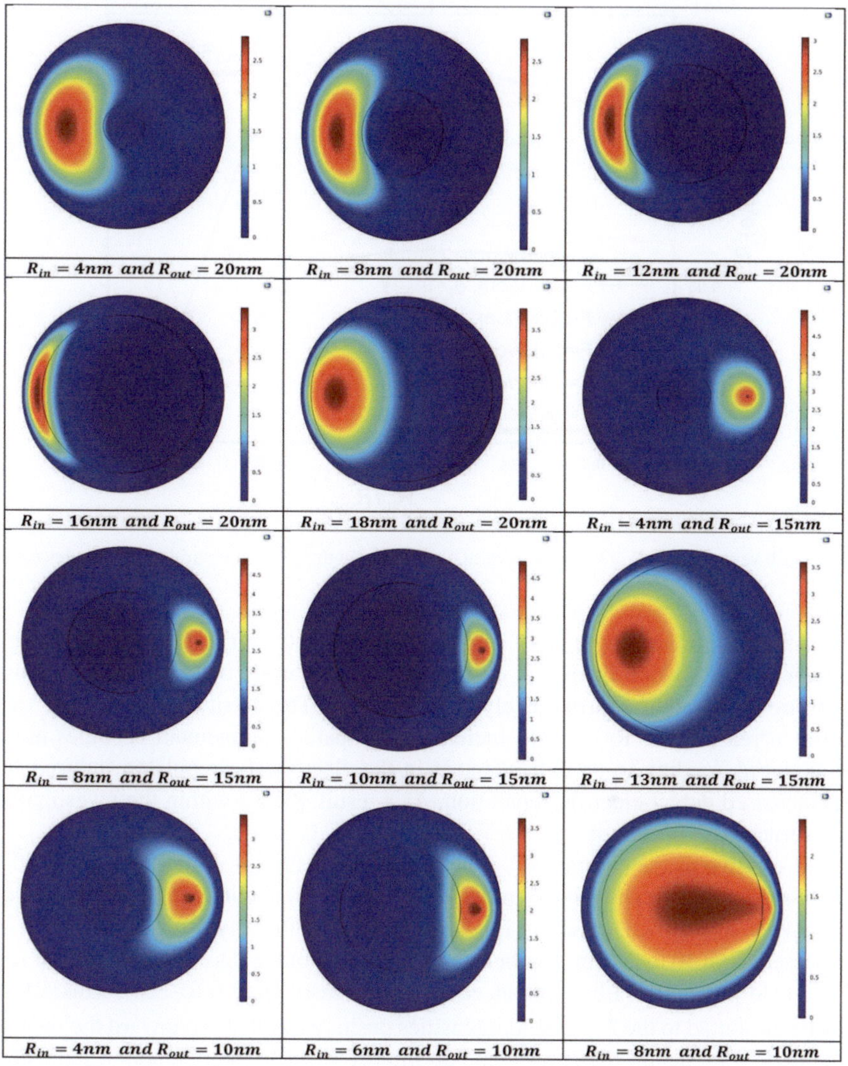

Fig. 2 Graph illustrating the electronic density as a function of nanostructure's size

electron. Furthermore, it is important to highlight that the peak binding energy is observed when both the φ_F and $\vec{F}_{(x,y)}$-field are set at 180° and 20 kV/cm, respectively. The specific azimuthal angle φ_F directs the $\vec{F}_{(x,y)}$-field opposite to $\varphi_F = 180°$, inducing a closer proximity between the electron and the defect, resulting in a confined spatial region. As a result, the ionization energy is heightened due to the enhanced electrostatic repulsion between the defect and the electron.

In a similar manner, Fig. 4 illustrates the evolution of the ground state defect Eb versus the ratio (R_{in}/R_{out}) for several Core's radii (R_{in}), taken in the consideration the

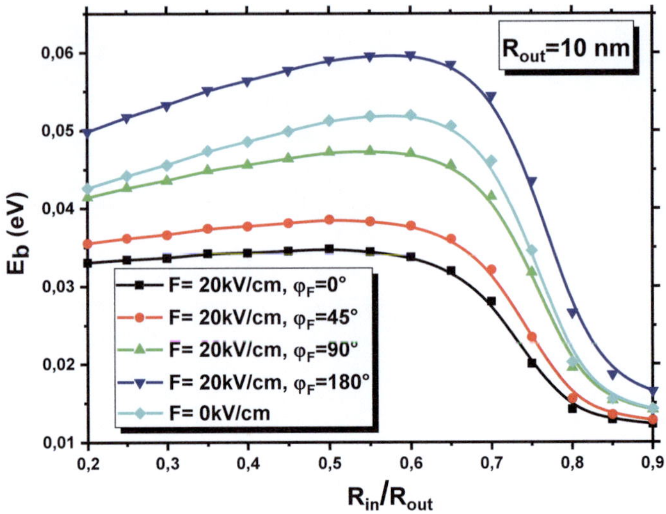

Fig. 3 Eb versus ratio (R_{in}/R_{out}) for several electric strengths (F) and its orientation angles (φ_F)

$\vec{F}_{(x,y)}$-field alignment angle (φ_F). The plot shows a behavior of the binding energy similar to that studied in Fig. 3, even though in this case, two dissimilar specific radii are observed at approximately 0.65 and 0.82. These critical points arise due to the increase in the R_{in}. As R_{in} increases, the spatial confinement becomes more pronounced, leading to an enhancement in the electrostatic repulsion induced by the increased defect-electron separation. As a result, the Eb within the investigated nanostructure is improved.

Irrespective of the external $\vec{F}_{(x,y)}$-field intensities, for $\frac{R_{in}}{R_{out}} \sim 0.2$ and $\varphi_F = 90°$, the binding energy exhibits a significant increase from about 0.02 eV (cyan curve) to 0.04 eV (red curve), resulting in an enhancement of nearly 50% when the R_{in} is increased from 10 to 20 nm. Furthermore, regardless of the disk's dimensions or the $\vec{F}_{(x,y)}$-field intensities, Eb is enhanced as φ_F increases from 0 to 180°. This effect can be attributed to the $\vec{F}_{(x,y)}$-field intensities pushing the electron toward the defect, leading to a reduction in distance and an improvement in the interaction repulsion between them. Consequently, the binding energy can only rise. Another notable observation is that the binding energy improves with an increase in the R_{out}, which is anticipated. The augmentation of the R_{out} intensifies geometrical confinement, leading to a decrease in the defect-electron separation, consequently leading to an increase in the ionization energy. Overall, the comportment of Eb changes notably within the interval of $\frac{R_{in}}{R_{out}}$ ranging from 0.6 to 0.8, corresponding to the boundary of a Quantum Well Wire system. In this range, the binding energy varies between 0.005 and 0.06 eV.

In Fig. 5, the Eb is presented versus the ratio (R_{in}/R_{out}) for various Shell radii (R_{out}) along with the external $\vec{F}_{(x,y)}$-field. The comportment of the binding energy as a function of R_{in}/R_{out} is found to be relatively comparable to that shown in Figs. 3

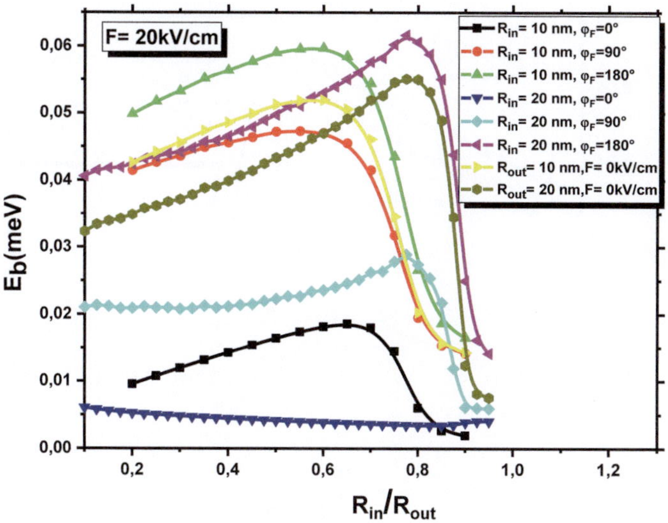

Fig. 4 Eb versus (R_{in}/R_{out}) for several R_{in} taken account the azimuthal angle of the $\vec{F}_{(x,y)}$-field

and 4, as discussed previously. Moreover, as depicted in Fig. 5, the behavior of the binding energy (BE) is significantly influenced by both the $\vec{F}_{(x,y)}$-field and the R_{out}. When maintaining a constant $\vec{F}_{(x,y)}$-field intensity (F = 0 kV/cm), augmenting the outer radius size (R_{out}) from 10 to 20 nm results in an increase in the binding energy. This is because the increase in R_{out} results in an improvement in the defect-electron separation, leading to enhanced electrostatic forces and, consequently, a growth in the ionization energy of the electron. However, irrespective of the system's dimensions, it is noted that the Eb diminishes with the escalation of the applied electric field intensity, as anticipated. This phenomenon can be rationalized by the electron's increased displacement from the impurity due to the higher electric field, leading to a heightened defect-electron separation and a decline in their electrostatic repulsion. Consequently, the binding energy weakens. For example, the binding energy is reduced by approximately 0.025 eV when considering $R_{in}/R_{out} \sim 0.2$, $R_{out} = 10\,\text{nm}$, and varying F from 0 to 20 kV/cm. This reduction becomes more prominent with increasing the ratio R_{in}/R_{out}, reaching around 0.7.

In Fig. 6 plots (a, b), the evolution of Eb with respect to the $\vec{F}_{(x,y)}$-field direction is depicted, considering the influence of the Shell radius (R_{out}). From the data presented in plot (a) of this graph, it becomes apparent that the Eb demonstrates a gradual augmentation when the $\vec{F}_{(x,y)}$-field aligns with the x-axis, specifically for φ_F values less than 50°. However, this enhancement accelerates notably for φ_F values exceeding 50° and culminates in a consistent value at $\varphi_F = 80°$, where the $\vec{F}_{(x,y)}$-field is nearly opposite to the x-axis. This comportment can be elucidated by the alteration in $\vec{F}_{(x,y)}$-field orientation, which shifts the electron in the direction of the defect, leading to a decrease in the defect-electron separation. Consequently, the electrostatic force between them intensifies. As a outcome, the Eb enhances. This interpretation is

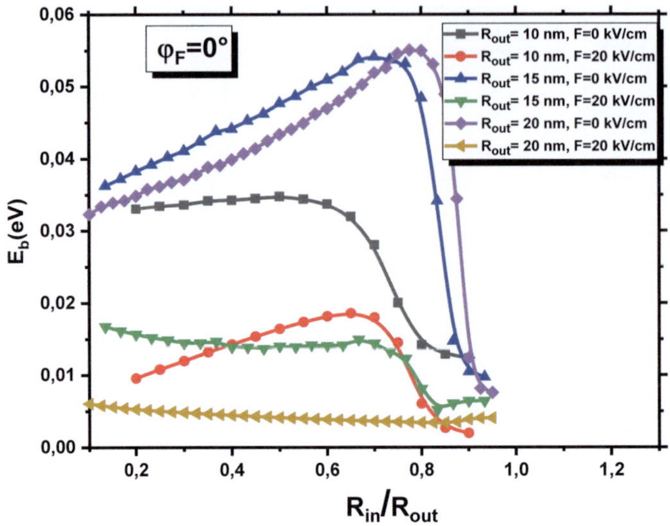

Fig. 5 Eb versus (R_{in}/R_{out}) for several R_{out} taken into account the external $\vec{F}_{(x,y)}$-field

further reinforced by the outcomes depicted in board (b). It is evident that nevertheless of the nanostructure's dimensions, the binding energy is significantly higher for φ_F about 180° compared to φ_F near 0°. In other words, the binding energy is considerable when the $\vec{F}_{(x,y)}$-field is oriented close to the opposite of the x-axis (− x).

Furthermore, it is clearly shown that the E_b is falls gradually with a growth in the outer diameter magnitude, as depicted in both panels (a, b). This effect occurs because as the outer radius (R_{out}) increases, the defect-electron separation becomes larger, resulting to a decrease in the Columbian interaction between them. Consequently, the binding energy gradually decreases. For instance, when considering a

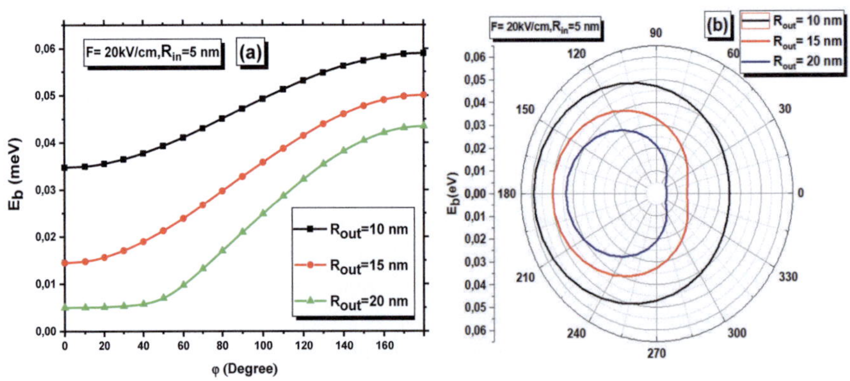

Fig. 6 Eb versus (φ_F) in dissimilar plots for several R_{out} with a given $R_{in} = 5$ nm

given $\vec{F}_{(x,y)}$-field strength (F = 20 kV/cm) and φ = 0° with R_{in} = 5 nm, the binding energy has shown a notable reduction near 0.015 eV (black curve) to 0.035 eV (red one), representing a difference of approximately 0.020 eV with the increase in the outer radius (R_{out}) from 10 to 15 nm. This reduction is attributed to the decrease in the defect-electron ionization energy thanks to the larger defect-electron separation. These results align significantly with the findings previously reported in Refs., indicating a consistent agreement with the existing literature [27, 28].

We have plotted the fluctuation of Eb in proportion to the $\vec{F}_{(x,y)}$-field strength in Fig. 7, shown in both graph (a) and (b), to analyze the influence of the external $\vec{F}_{(x,y)}$-field and its proper angle φ_F. With growing external $\vec{F}_{(x,y)}$-field intensity, it is clear from graph (a) that the Eb displays a linear fall until stabilizing at a fixed value associated with a critical $\vec{F}_{(x,y)}$-field strength. The value of the outside diameter (R_{out}) determines where the critical point lies. The decreasing electrostatic contact between the defect and the electron is responsible for the apparent decrease. The expansion of the electron wave function under the influence of the $\vec{F}_{(x,y)}$-field causes this phenomena. The responsiveness of the binding energy to the external $\vec{F}_{(x,y)}$-field drops beyond a certain critical threshold. This specific point changes as the outer radius grows, becoming smaller as R_{out} grows. Based on the direction of the $\vec{F}_{(x,y)}$-field, the Eb in graph (b) displays two dissimilar comportments. The Eb falls linearly with increasing $\vec{F}_{(x,y)}$-field intensity when $\varphi_F < 90°$ (positive direction, + x) (black, red, and green lines). In contrast, the binding energy grows linearly with the applied electric field intensity for $\varphi_F > 90°$ (negative direction, – x) (blue line). The behavior of the electron under various $\vec{F}_{(x,y)}$-field directions sheds light on this event. The electron moves away from the defect when the $\vec{F}_{(x,y)}$-field is positive (+ x), increasing the distance between them and resulting in a decrease in Eb. The electron is pulled toward the defect when the $\vec{F}_{(x,y)}$-field is directed negatively (– x), strengthening the electrostatic contact and raising the ionization energy. For example, for a given system's size (R_{out} = 20 *and* R_{in} = 10 nm) and an electric field intensity of approximately 50 kV/cm, The ionization energy has increased significantly from roughly 0.065 eV (blue line) to 0.01 eV (black line), signifying a reduction of approximately 0.055 eV by modifying the $\vec{F}_{(x,y)}$-field direction from φ_F: 180 to 0°. Overall, it is evident that the Eb is strongly influenced by the $\vec{F}_{(x,y)}$-field intensity and its direction. These conclusions about the direction and magnitude of the external $\vec{F}_{(x,y)}$-field are in good agreement with those that were previously published for a cylindrical quantum dot (CQD) in Ref. [9].

3.2 *The Stark-Shifts*

In Fig. 8, panels (a, b), we have illustrated the Stark-Shift (ΔE_{SS}) versus the radius ratio (R_{in}/R_{out}) for different electric field directions (φ_F) while keeping the Shell (panel a) and Core (panel b) radius fixed. Notably, from panel (a), ΔE_{SS} exhibits two

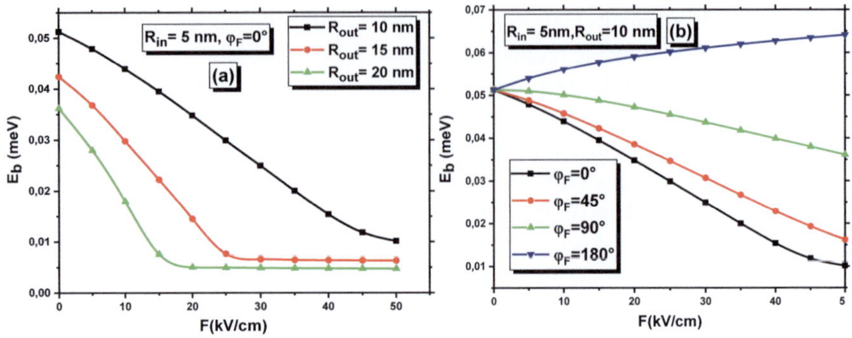

Fig. 7 E_b versus the external $\vec{F}_{(x,y)}$-field for several R_{out} (**a**) and different φ_F (**b**) with a given $R_{in} = 5$ nm

distinct regimes and displays a significant similarity to the comportment of ionization energy. It decreases smoothly concerning the ratio (R_{in}/R_{out}) until reaching a smallest value at a specific point of approximately 0.65. Beyond this critical point, ΔE_{SS} grows rapidly and attains a steady region for values above 0.9, where R_{in} and R_{out} remains nearly equivalent to each other. This behavior can be explained regardless of the $\vec{F}_{(x,y)}$-field intensity or φ_F and for a given $R_{out} = 10$ nm. When R_{in} rises, the defect-electron separation also increases, leading to an enhancement in the Columbian interaction between them, resulting in a drop in ΔE_{SS}. Conversely, when R_{in} approaches or surpasses R_{out}, the defect-electron distance decreases, and the electronic wave-function becomes more confined about the defect thanks to geometrical confinement. As a result, ΔE_{SS} remains stronger and growths quickly.

It is clear that ΔE_{SS} rises with the $\vec{F}_{(x,y)}$-field direction (φ_F) ranging from 0 to 180°, and there is a change in behavior for $\varphi_F > 90°$. These outcomes are reliable and in agreement with those mentioned in Fig. 9. Figure 9 presents the variation of Stark-Shift (ΔE_{SS}) as a function of R_{in}/R_{out} for different orientation angles as well

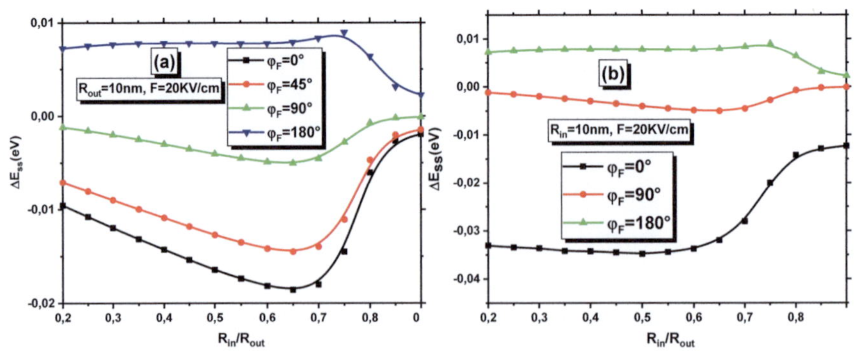

Fig. 8 ΔE_{SS} versus (R_{in}/R_{out}) with several $\vec{F}_{(x,y)}$-field angles (φ_F) for given, $R_{out} = 10$ nm (**a**), $R_{in} = 10$ nm (**b**) with $F = 20$ kV/cm

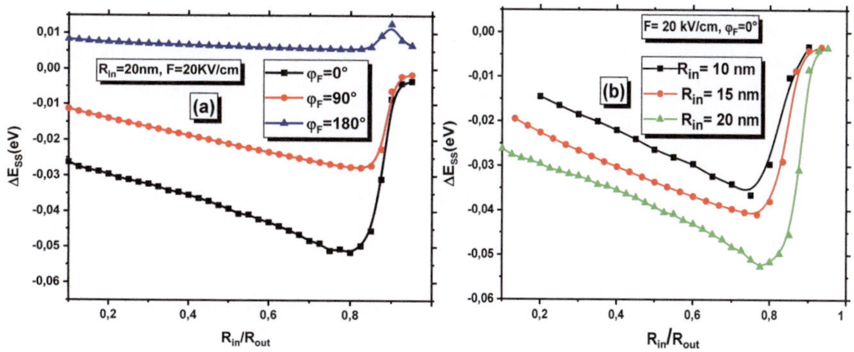

Fig. 9 ΔE_{SS} versus (R_{in}/R_{out}) with several $\vec{F}_{(x,y)}$-field angles (φ_F), plot (**a**) and dissimilar R_{in}, plot (**b**)

as R_{in} values. It is evident that ΔE_{SS} strongly influenced by the system's dimensions and the electric field orientation (φ_F). ΔE_{SS} enhances with growing φ_F whereas it decreases with increasing the magnitude of R_{in}.

The observed behaviors in Fig. 8 regarding the Stark-Shift are consistent with the explanations provided earlier, considering the effects of $\vec{F}_{(x,y)}$-field, direction, and size. However, for a fixed outer radius ($R_{out} = 20\,\text{nm}$) and $\vec{F}_{(x,y)}$-field intensity (F = 20 kV/cm), it is notable from graph (a) that the ΔE_{SS} has shifted from nearly − 0.027 eV (black curve) to 0.009 eV (blue curve), This modification results in a change in direction from 0 to 180° that causes a shift of around 0.117 eV. This improvement is especially noticeable for lower outer radius values that match to (i.e., $R_{in}/R_{out} \sim 0.8$). Additionally, as shown in panel (b), it is evident that when the ratio gets closer to 1 (i.e., $R_{in} \sim R_{out}$), the Stark-Shift almost stops responding to changes in the inner radius. In other words, the Stark-Shift remains relatively constant when the ratio R_{in}/R_{out} approaches unity, indicating a saturation of the Stark-Shift effect in this region.

To gain an overview of the $\vec{F}_{(x,y)}$-field's impact on the ΔE_{SS}, we depicted Fig. 10, illustrating the Stark-Shift concerning the external $\vec{F}_{(x,y)}$-field, counting the influence of the φ_F. Evidently, the Stark-Shift exhibits a similar behavior to the binding energy (BE), as discussed in plot (b) of Fig. 7. The fundamental underlying factors behind these comportments can be deemed analogous to those affecting the binding energy, elucidated in the mentioned figure. Moreover, this similarity can also be attributed to the relationship that interconnects these quantities, as mentioned in Eqs. (20) and (21). Nevertheless, Fig. 11 displays the evolution of $|\Delta E_{SS}|$ concerning the external $\vec{F}_{(x,y)}$-field intensity, considering the impact of the outside diameter change. Irrespective of the inner radius and electric field direction, $|\Delta E_{SS}|$ increases with the external $\vec{F}_{(x,y)}$-field intensity pending it touches a constant regime, where $|\Delta E_{SS}|$ rests stable for all $\vec{F}_{(x,y)}$-field strengths. This stable regime is predominantly attained at a critical point of the $\vec{F}_{(x,y)}$-field, contingent on the outside diameter. As the outer radius size increases, the critical value diminishes. These findings harmonize well with the results previously reported in panel (a) of Fig. 7, which pertain to the binding energy as a function of $\vec{F}_{(x,y)}$-field intensity and the outside diameter. The physical factors influencing $|\Delta E_{SS}|$ under the $\vec{F}_{(x,y)}$-field are akin to those affecting the binding energy in the examined nanostructure.

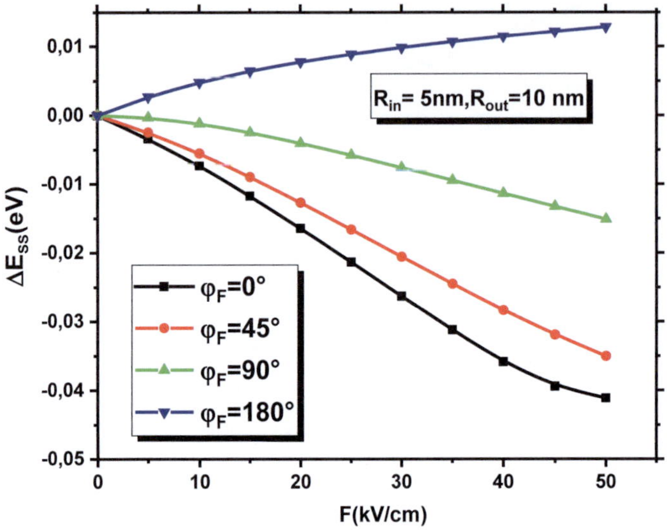

Fig. 10 ΔE_{SS} versus $\vec{F}_{(x,y)}$-field intensity for various φ_F

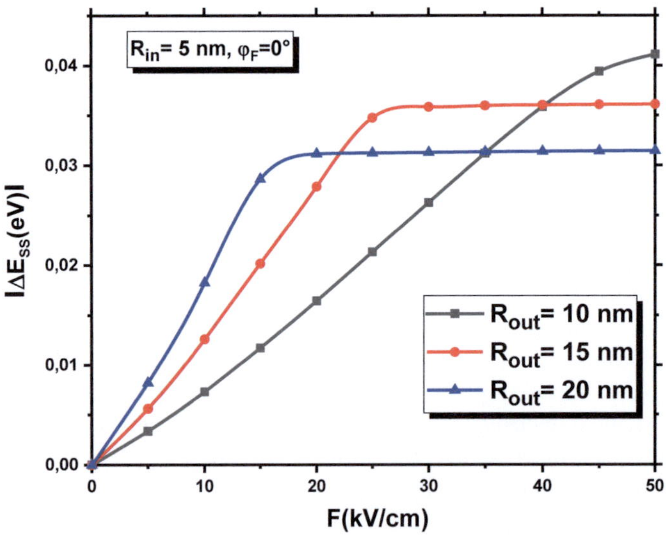

Fig. 11 $|\Delta E_{SS}|$ versus the external $\vec{F}_{(x,y)}$-field intensity for dissimilar R_{out}

Alternately, the size of the shell and the core has an impact on the Stark effect. The amplitude of the Stark shift lowers (improves in absolute value) as the shell radius rises while the applied electric field is below the critical threshold. However, this effect often breaks down at electric fields greater than 30 kV/cm. This finding may be helpful for electro-modulator devices, which regulate a light beam's intensity in response to a certain applied electric field.

3.3 The Diamagnetic Response

In Fig. 12 plots (a, b, c), the magnetic susceptibility χ_{Dia} is presented dependent on the $\vec{F}_{(x,y)}$-field strength for several R_{out} magnitude, taken into account diverse φ_F ranging from the x-axis to the opposite one, with a fixed R_{in} value of 5 nm. It is evident from these graphs that, for fixed R_{in}, the χ_{Dia} becomes more pronounced as the outside diameter (R_{out}) decreases. This effect is attributed to the stronger confinement experienced by the electrons at smaller R_{out} values. Additionally, it is evident that the external $\vec{F}_{(x,y)}$-field induces motion of the electrons in the opposite direction of $\vec{F}_{(x,y)}$.

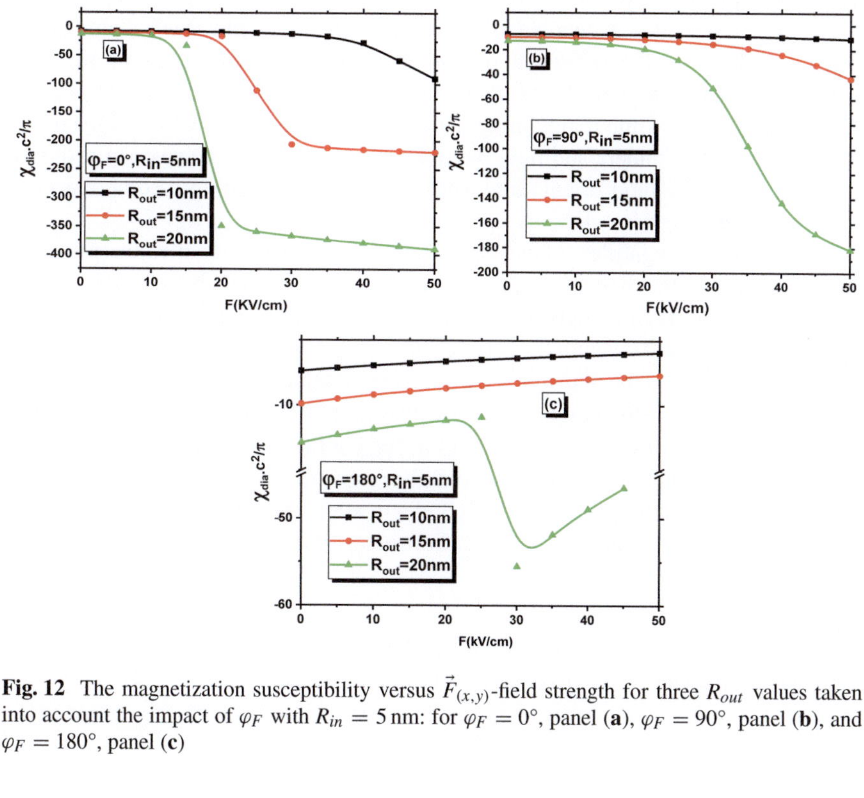

Fig. 12 The magnetization susceptibility versus $\vec{F}_{(x,y)}$-field strength for three R_{out} values taken into account the impact of φ_F with $R_{in} = 5\,\text{nm}$: for $\varphi_F = 0°$, panel (**a**), $\varphi_F = 90°$, panel (**b**), and $\varphi_F = 180°$, panel (**c**)

It is also clear from panel (c) of the same picture that the χ_{Dia} falls as the $\vec{F}_{(x,y)}$-field intensity increases until a critical threshold, at which point it stabilizes. It is noteworthy that as the external $\vec{F}_{(x,y)}$-field direction changes from the x to the y-direction, this critical threshold moves towards greater values. The patterns in panels (a) and (b), however, differ from those in panel (c). This variation is due to the fact that when φ_F is equal to 180°, the defect-electron interaction becomes stronger, resulting in increased localization of the electronic wave-function and a stronger χ_{Dia}. Specifically, for the case of $R_{out} = 20$ nm, a peak is observed around F = 30 kV/cm, corresponding to tunneling behavior that reduces geometrical confinement, The χ_{Dia} is reduced as a result of this activity. However, it is important to note that increasing the $\vec{F}_{(x,y)}$-field results in a reduction in the distance between the defect and the electron, which increases the magnetization susceptibility. In addition, at $\vec{F}_{(x,y)}$-field of F < 10 kV/cm, the χ_{Dia} is comparatively steady when the $\vec{F}_{(x,y)}$-field is directed between the x and y-directions. To be more precise, it maintains a value of $\chi_{Dia} \cdot c^2/\pi = -10.3480$ for different shell radius sizes. This equilibrium is attained because the effects of the electric field and the geometric system within this range are in equilibrium.

Our numerical outcomes show striking agreement with the conclusions made for a spherical quantum box in the research of Rezaei et al. [29]. Table 1 clearly illustrates that χ_{Dia} dramatically decreases with increasing shell radius (R_{out}) while it improves with an rise in the external $\vec{F}_{(x,y)}$-field.

Figure 13 provides a detailed illustration of the χ_{Dia} and sheds light on the physics underlying the influence of φ_F. It presents the evolution of the χ_{Dia} concerning the direction of the $\vec{F}_{(x,y)}$-field φ_F for several shell radii, namely $R_{out} =$ 10, 15, and 20 nm, while maintaining a fixed core radius at $R_{in} = 5$ nm. This figure allows us to better understand how the diamagnetic response behaves in response to different electric field orientations and shell sizes, facilitating a comprehensive analysis of the system's behavior.

It is evident that regardless of the core radius, an growth in the shell radius produces to a diminution in the diamagnetic response. This behavior may be explained by the larger dimension of the nanostructure, which results in less confinement because of the geometry and eventually reduces the χ_{Dia}. Moreover, the impact of the azimuth angle (φ_F) on the diamagnetic response is notable. The diamagnetic response enhances as the $\vec{F}_{(x,y)}$-field moves from the positive to the negative side of the x-axis, culminating at $\varphi_F = 180°$. This happens as a result of the $\vec{F}_{(x,y)}$-field pulling electrons toward an defect and forcing them into a smaller area, thereby enhancing their contact with the defect (resulting in an enhanced electrostatic attraction) and consequently improving the magnetization properties. Conversely, the diamagnetic response starts

Table 1 Some numerical values of χ_{Dia} for several shell diameter (R_{out}) for $\varphi_F = 180°$

R_{out} (nm)	10	15	20
F (kV/cm)	50	30	20
$\chi_{Dia} \cdot c^2/\pi$	– 93.0627	– 224.1105	– 383.2519

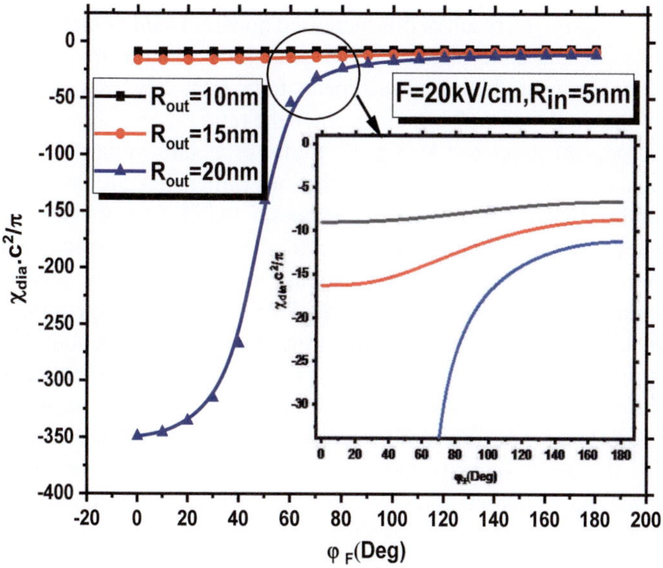

Fig. 13 χ_{Dia} versus φ_F for different R_{out} with $R_{in} = 5\,\text{nm}$ and F = 20 kV/cm

with a minimum at $\varphi_F = 0°$, primarily due to the $\vec{F}_{(x,y)}$-field increasing the defect-electron separation when directed in the x-direction. Furthermore, it is observed that the effect of φ_F is more pronounced in wider quantum dots ($R_{out} = 20\,\text{nm}$) compared to narrower ones ($R_{out} = 10\,\text{nm}$), as illustrated in the sub-plot.

Table 2 provides some numerical values of χ_{Dia} for dissimilar magnitudes of the outside diameter (R_{out}) with various electric field orientations, showcasing the dependence of the diamagnetic response on these parameters.

In Fig. 14, we have depicted the evolution of the magnetization properties versus the ratio R_{in}/R_{out}, for several magnitudes of R_{out} with an $\vec{F}_{(x,y)}$-field intensity of 20 kV/cm for various orientations: $\varphi_F = 0°$ in panel (a), $\varphi_F = 90°$ in panel (b), and $\varphi_F = 180°$ in panel (c). From these panels, as the ratio of inner radius (R_{in}) to outer radius (R_{out}) increases from 0 to 1, the observed pattern reveals a gradual increase in magnetization properties. The behavior that results in a stronger confinement zone for the defect and electron when the R_{in}/R_{out} ratio increases while the inner radius remains constant provides an explanation for this phenomenon. The increased

Table 2 Some numerical values of χ_{Dia} for dissimilar magnitudes of the outside diameter (R_{out}) with different electric field orientations

$R_{in} = 5\,\text{nm}$	$R_{out} = 10\,\text{nm}$	$R_{out} = 15\,\text{nm}$	$R_{out} = 20\,\text{nm}$
$\vec{F}$ direction	$\chi_{Dia} \cdot c^2/\pi$		
$\varphi_F = 0°$	− 9.04848	− 16.29905	− 349.10061
$\varphi_F = 90°$	− 7.82349	− 11.7929	− 19.4464
$\varphi_F = 180°$	− 6.60463	− 8.69761	− 11.15924

diamagnetic response is a result of this increased quantum confinement, which results from the shape of the nanostructure. This behavior is in agreement with previous literature [24]. However, the response appears to be relatively insensitive to changes in the size of the shell radius (R_{out}). Thus, when the ratio R_{in}/R_{out} approaches 1 ($R_{in} \sim R_{out}$), the diamagnetic response decreases rapidly, especially around $R_{in}/R_{out} = 80\%$, approaching a critical value associated with a one dimension circular QWW. At this boundary comportment, the electron undergoes a switch from the shell material to the core one, causing a reduction in the electrostatic interaction between the defect and the electron and leading to a rapid decrease in the diamagnetic response.

In particular, for $R_{out} = 20$ nm in panel (c), Due to the interaction of the $\vec{F}_{(x,y)}$-field and geometric restrictions, the magnetization characteristic gradually diminishes. Particularly in comparison to other outer radii, there is clearly noticeable strong confinement at an outside radius of $R_{out} = 20$ nm. In this case, as seen in Fig. 1, the $\vec{F}_{(x,y)}$-field's impact becomes more prominent. Since a result, the magnetization characteristic decreases since the electrons have a tendency to go in the direction of the barrier material. This phenomenon also impacts the donor dipole moment and

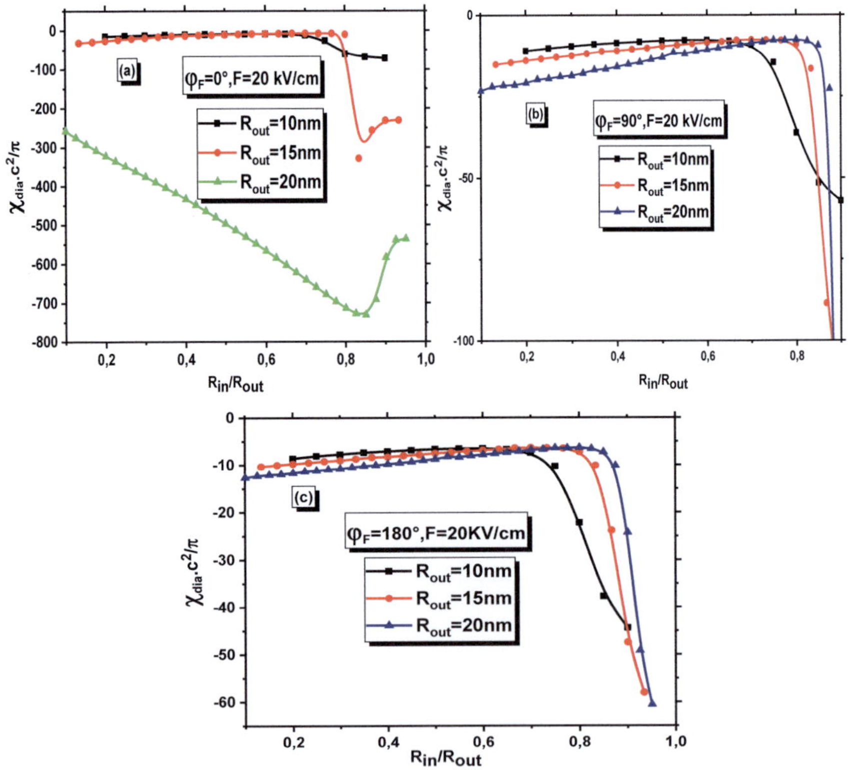

Fig. 14 χ_{Dia} versus R_{in}/R_{out} for several magnitudes of R_{out} with $F = 20$ kV/cm for different orientations: $\varphi_F = 0°$, panel (**a**), $\varphi_F = 90°$, panel (**b**), and $\varphi_F = 180°$, panel (**c**)

the polarizability thanks to their interrelation, as will be discussed next. Additionally, as the R_{in}/R_{out} ratio increases, the electrons experience high restriction, making it easier for them to overcome the barrier material.

The diamagnetic susceptibility of this system can be tailored for specific size parameters, which are defined by the strength of the applied electric field and the ratio of the inner diameter (R_{in}) to the outside diameter (R_{out}), as shown in Figs. 12, 13, and 14. This characteristic offers the opportunity to design optoelectronic and photonics devices that leverage the magnetization susceptibility of the behavior to achieve desired functionalities and performance. By carefully selecting the size parameters and electric field strength, one can fine-tune the diamagnetic response, enabling the creation of versatile and efficient optoelectronic devices for various applications.

3.4 The Polarizability

Figure 15 illustrates the fluctuation of the polarizability concerning the proportion R_{in}/R_{out}, taken into the consideration the impact of R_{out} with an applied electric field strength of 20 kV/cm for various orientations, assuming an on-center impurity. The direction of the electric field, φ_F, largely determines the trend of the polarizability with respect to size. Panels (a) and (b) show that for all (R_{in}/R_{out}) values below 70%, the polarizability gradually increases, with its scale approximately ranging from $|\alpha_P| \sim [0.01684, 1.548] \cdot 10^{-12}\,\mu\text{F}\,\text{nm}^2$. This phenomenon is ascribed to the heightened geometric restriction, whereby alterations in the electronic wave function characteristics influence the polarizability. This suggests a notable shift in the electron wave function localization. However, when the (R_{in}/R_{out}) ratio approaches to the unity, the polarizability experiences a rapid decrease before exhibiting an increase again, due to the tunnel effect.

This pattern can be elucidated by the phenomenon of shell electrons undergoing tunneling into barrier material when the R_{in}/R_{out} values exceed 0.8. This eventuality leads to a reduction in the polarizability. Furthermore, the strengthening of confinement induced by the structure's dimension within the shell area is associated with the electron approaching the defect, resulting in an amplified polarizability. Conversely, the polarizability displays a contrasting trend in Panel (c) of the same figure, because of the $\vec{F}_{(x,y)}$-field being positioned in the reverse direction relative to the x-direction ($\varphi_F = 180°$). As the external $\vec{F}_{(x,y)}$-field intensity compels the electrons to move counter to the external investigated field, influenced by the Stark effect, it results in a decrease in the polarizability value at the peak summit. Additionally, the location of the peak shifts to higher values of (R_{in}/R_{out}) as R_{out} increases, which is logical since the tunneling effect is more prominent for thinner systems ($R_{out} = 10$ nm) compared to wider ones ($R_{out} = 20$ nm).

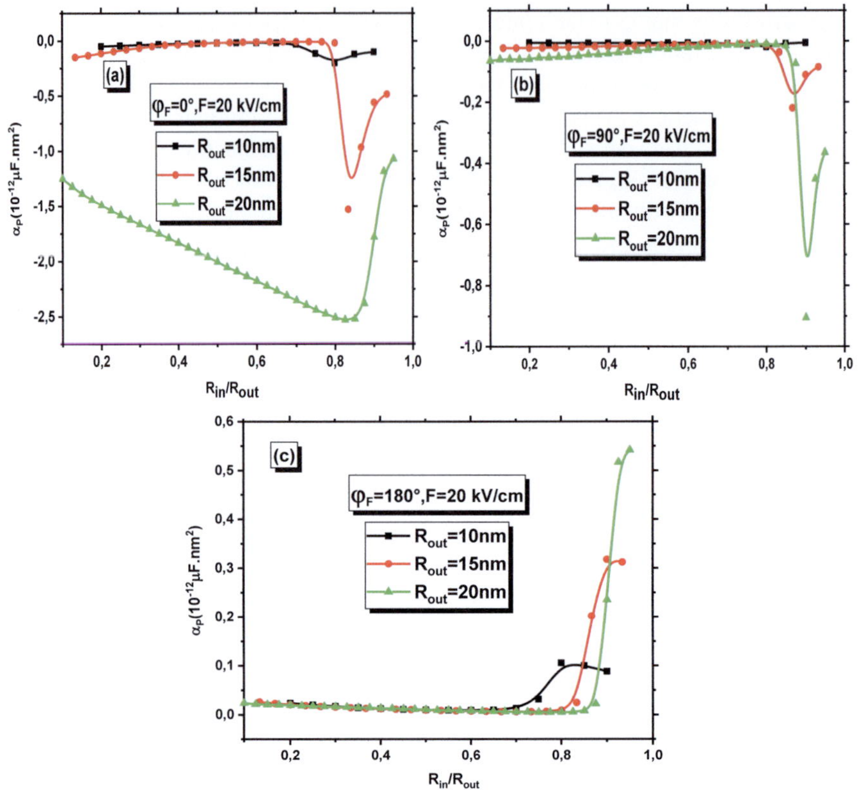

Fig. 15 The polarizability versus R_{in}/R_{out} taken into the consideration the effect of R_{out} with $F = 20$ kV/cm for different orientations: $\varphi_F = 0°$, panel (**a**), $\varphi_F = 90°$, panel (**b**), and $\varphi_F = 180°$, panel (**c**)

3.5 *The Dipole Moment*

In Fig. 16 plots (a, b, c), we illustrate the dipole moment's fluctuation concerning the ratio R_{in}/R_{out} taken into the consideration the impact of R_{out} with F = 20 kV/cm and different electric field orientations, considering an on-center defect. As anticipated, the dipole moment's behavior closely mirrors that of the polarizability, as indicated by Eq. (15), which clearly establishes their linear relationship. Consequently, the patterns seen in the pictures may be examined using the same criteria used to evaluate Fig. 15. Furthermore, for $0.1 < R_{in}/R_{out} < 0.8$ tiny system dimensions, The dipole moment demonstrates a linear dependence on R_{in}/R_{out}, particularly in the regimes of robust and low confinement. This suggests a heightened manifestation of the linear Stark effect within these confinement scenarios. However, for $R_{in}/R_{out} > 0.8$, corresponding to very strong quantum confinement, the dipole moment's behavior changes Because of the tunneling phenomenon, the stark shift becomes notably

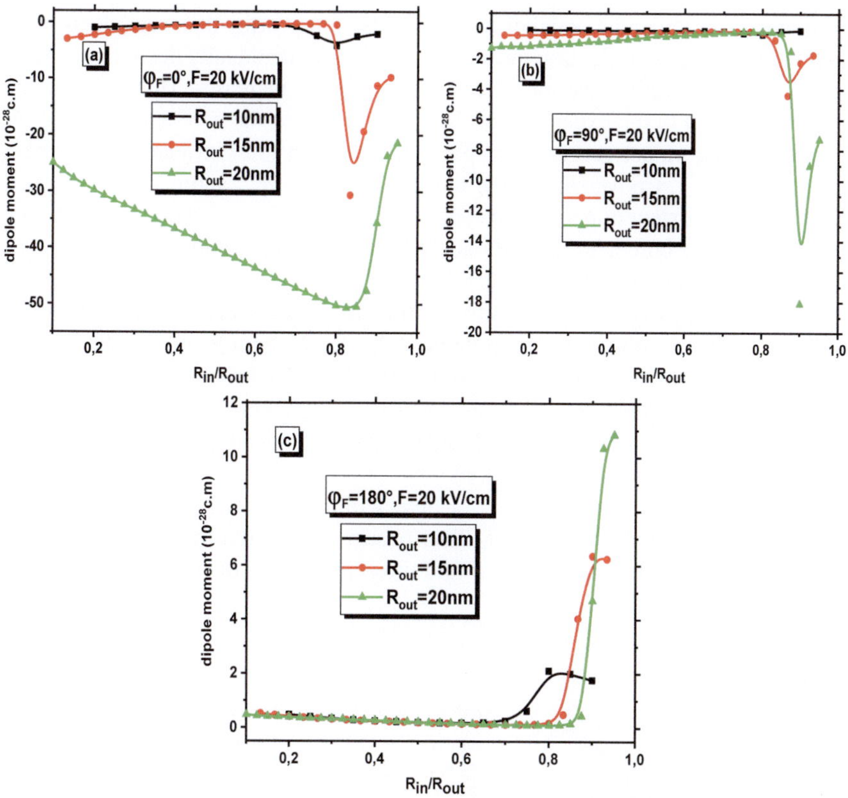

Fig. 16 The fluctuation of dipole moment versus the portion R_{in}/R_{out} taken into the consideration the impact of R_{out} with $F = 20$ kV/cm for different orientations: $\varphi_F = 0°$, panel (**a**), $\varphi_F = 90°$, panel (**b**), and $\varphi_F = 180°$, panel (**c**)

impactful within this range, exerting a substantial influence on the dipole moment's behavior.

3.6 *The Electron-Impurity Distance*

To explore the impact of the system sizes and external $\vec{F}_{(x,y)}$-field on the defect-electron separation, we analyze the fluctuation of the average defect-electron separation (r_{De}) with respect to the ratio (R_{in}/R_{out}), the $\vec{F}_{(x,y)}$-field intensity, and its proper angle (φ_F) in the plots (a, b, c) of Fig. 17. Except for panel (c), the figure shows that r_{De} increases steadily with the $\vec{F}_{(x,y)}$-field intensity. This occurs because the external $\vec{F}_{(x,y)}$-field drives the electrons towards the donor defect, leading to a reduction in the defect-electron average separation. Furthermore, the impact of the investigated field is more prominent for bigger quantum dots (QDs) with an outer radius (R_{out}

= 20 nm), as compared to smaller ones. It is also noteworthy that, regardless of the shell radius, the defect-electron mean square separation gradually decreases initially until it touches a minimum point, after which it quickly increases. This phenomenon can be attributed to the expansion of the region between the core and shell regions for a given R_{out} value, causing the defect and electron to experience confinement, thus reducing r_{De}. Conversely, when the ratio of R_{in} to R_{out} approaches unity (i.e., $R_{in} \sim R_{out}$), the extent of geometric confinement maximizes, causing the electron to shift closer to the core material, ultimately leading to an elevation in r_{De}.

Table 3 presents the values of the defect-electron separation with and without the external $\vec{F}_{(x,y)}$-field across various confinement conditions. The data presented in the table clearly illustrates the notable influence of both geometric factors and external $\vec{F}_{(x,y)}$-field on the defect-electron separation. As the shell's diameter (R_{out}) and external $\vec{F}_{(x,y)}$-field intensity increase, there is a corresponding increase in this distance. This effect can be attributed to the heightened quantum confinement resulting from these specific conditions.

The influence of the external $\vec{F}_{(x,y)}$-field within the disk's geometry serves to diminish the confinement's impact. This observation is reinforced by the agreement between the findings in Table 4 and those previously documented in Table 3,

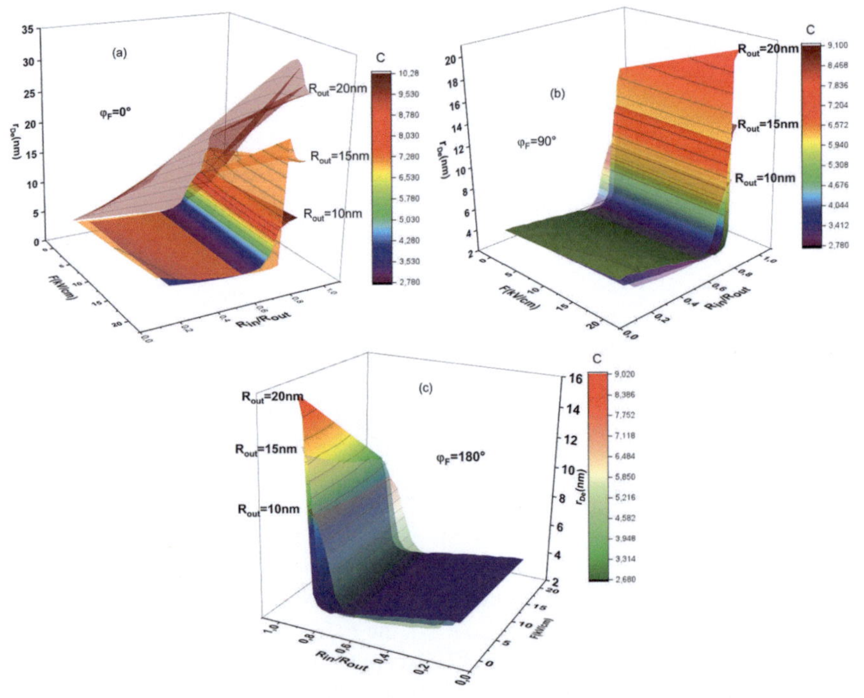

Fig. 17 The defect-electron separation versus of $\vec{F}_{(x,y)}$-field (x-direction) and ratio R_{in}/R_{out} (y-direction) for several magnitudes of R_{out} with different $\vec{F}_{(x,y)}$-field orientations: $\varphi_F = 0°$, panel (**a**), $\varphi_F = 90°$, panel (**b**), and $\varphi_F = 180°$, panel (**c**)

Table 3 The impact of sizes and $\vec{F}_{(x,y)}$-field on the defect-electron separation (r_{De}) in two different confinement regimes

$\varphi_F = 90°$		Without electric field $F = 0$ kV/cm	With strong electric field $F = 20$ kV/cm
R_{out} (nm)	Confinement regime	r_{De} (nm)	
10	Weak $\frac{R_{in}}{R_{out}} \sim 0, 1$	3.4078	3.4729
15		3.7798	4.0625
20		4.0653	4.7737
10	Strong $\frac{R_{in}}{R_{out}} \gg 0, 1$	2.799	2.8724
15		2.7255	2.8349
20		2.6277	2.7284

Table 4 The influence of the proper angle of $\vec{F}_{(x,y)}$-field on the defect-electron separation (r_{De})

$F = 20$ kV/cm		$\varphi_F = 0°$	$\varphi_F = 90°$	$\varphi_F = 180°$
R_{out} (nm)	Confinement regime	r_{De} (nm)		
10	Weak $\frac{R_{in}}{R_{out}} \sim 0, 1$	4.0115	3.4729	3.1102
15		5.6374	4.0625	3.4819
20		19.771	4.7737	3.7929
10	Strong $\frac{R_{in}}{R_{out}} \gg 0, 1$	3.0081	2.8724	2.6902
15		2.8328	2.8349	2.6377
20		29.261	2.7284	2.7389

particularly regarding the influence of (R_{out}). Furthermore, a noteworthy observation is that altering the orientation from positive to negative direction leads to a decrease in the defect-electron separation, particularly pronounced in the regime of weak quantum confinement $\left(\frac{R_{in}}{R_{out}} \sim 0, 1\right)$. These outcomes align harmoniously with previously reported literature [30].

4 Conclusions

In the first section, we conducted a comprehensive investigation of the ionization energy and the Stark-shift in GaAs/GaAlAs homogenous quantum disk, taking into account the influence of sizes' nanostructure and external $\vec{F}_{(x,y)}$-field. The Schrodinger equation was solved within the effective-mass approximation using finite element analysis with proper consideration of the conditions at the interface separating the Core and Shell areas. Our findings demonstrate that both the direction of the external $\vec{F}_{(x,y)}$-field and the nanostructure sizes significantly affect the ionization energy and Stark-shift.

Specifically, we observed a decrease in both the ionization energy and Stark-shift under the external $\vec{F}_{(x,y)}$-field. However, this behavior could be modulated by regulating the system's dimensions, including the outer and inner diameters, as well as the proper angle of the $\vec{F}_{(x,y)}$-field. Notably, we found that the ionization energy and Stark-shift show significant variations with changes in the proper angle of $\vec{F}_{(x,y)}$-field, especially from 0 to 180°. These outcomes provide valuable insights and complement previous theoretical studies focused on the ionization energy and Stark-shift in our investigated system, such as quantum dots. Moreover, the results have practical implications for electro-modulator devices, as they offer the potential to control the intensity of a light beam by applying specific $\vec{F}_{(x,y)}$-field voltages in the interval of F from 0 to 50 kV/cm.

In the second section, we explored the impact of an external $\vec{F}_{(x,y)}$-field and the sizes of the core/shell quantum disks on the magnetization susceptibility, polarizability, and electric moment, using the same computational approach as in the first section. Our investigation revealed that these parameters play a crucial role in determining the physical properties of the quantum disks and can be optimized by adjusting the excitation conditions. This study delivers appreciated insights for the development of advanced GaAs-based photonic devices, where precise control over these physical properties is essential for achieving desired functionalities. Overall, our research contributes to a better understanding of the behavior of inhomogeneous quantum system under $\vec{F}_{(x,y)}$-field and offers potential avenues for enhancing the performance of nanoelectronic devices.

References

1. A. Zounoubi, K.E. Messaoudi, I. Zorkani, A. Jorio, Magnetic field and finite barrier-height effects on the polarizability of a shallow donor in a GaAs quantum wire. Superlattices Microstruct. **30**(4), 189–200 (2001). https://doi.org/10.1006/spmi.2001.1006
2. B.K. Ghosh, M. Hasanuzzman, I. Saad, K.A. Mohamad, M.K. Hossain, Photovoltaic technologies photo-thermal challenges: thin active layer solar cells significance. Optik **274**, 170567 (2023). https://doi.org/10.1016/j.ijleo.2023.170567
3. A. Bera, S. Saha, J. Ganguly, M. Ghosh, Noise-driven diamagnetic susceptibility of impurity doped quantum dots: role of anisotropy, position-dependent effective mass and position-dependent dielectric screening function. Chem. Phys. **474**, 36–43 (2016). https://doi.org/10.1016/j.chemphys.2016.05.010
4. R. Ghosh Chaudhuri, S. Paria, Core/shell nanoparticles: classes, properties, synthesis mechanisms, characterization, and applications. Chem. Rev. **112**(4), 2373–2433 (2012). https://doi.org/10.1021/cr100449n
5. A.R. Kortan et al., Nucleation and growth of CdSe on ZnS quantum crystallite seeds, and vice versa, in inverse micelle media. J. Am. Chem. Soc. **112**(4), 1327–1332 (1990). https://doi.org/10.1021/ja00160a005
6. G. Bastard, Hydrogenic impurity states in a quantum well: a simple model. Phys. Rev. B **24**(8), 4714–4722 (1981). https://doi.org/10.1103/PhysRevB.24.4714
7. L. Brey, N.F. Johnson, B.I. Halperin, Optical and magneto-optical absorption in parabolic quantum wells. Phys. Rev. B **40**(15), 10647–10649 (1989). https://doi.org/10.1103/PhysRevB.40.10647

8. J. Hermanson, Exciton and impurity states in rare-gas solids. Phys. Rev. **150**(2), 660–669 (1966). https://doi.org/10.1103/PhysRev.150.660
9. C. Heyn, C.A. Duque, Donor impurity related optical and electronic properties of cylindrical GaAs-$Al_xGa_{1-x}As$ quantum dots under tilted electric and magnetic fields. Sci. Rep. **10**(1), 1–18 (2020)
10. C.A. Duque, N. Porras-Montenegro, Z. Barticevic, M. Pacheco, L.E. Oliveira, Electron-hole transitions in self-assembled InAs/GaAs quantum dots: effects of applied magnetic fields and hydrostatic pressure. Microelectron. J. **36**(3), 231–233 (2005). https://doi.org/10.1016/j.mejo.2005.04.001
11. M. Gussoni, M. Rui, G. Zerbi, Electronic and relaxation contribution to linear molecular polarizability. An analysis of the experimental values. J. Mol. Struct. **447**, 163–215 (1998). https://doi.org/10.1016/S0022-2860(97)00292-5
12. M.H. Hachemi, M. Benchehima, K. Bencherif, H. Abid, The effect of N-incorporation on the structural and optoelectronic properties of GaP and GaAs for optical telecommunication applications: first-principles study. Optik **262**, 169282 (2022). https://doi.org/10.1016/j.ijleo.2022.169282
13. E. Arici, D. Meissner, F. Schäffler, N.S. Sariciftci, Core/shell nanomaterials in photovoltaics. Int. J. Photoenergy **5**, 199–208. https://doi.org/10.1155/S1110662X03000333
14. V.A. Harutyunyan, K.S. Aramyan, H.S. Petrosyan, G.H. Demirjian, Optical transitions in spherical quantized layer under the presence of radial electrical field. Phys. E **24**(3), 173–177 (2004). https://doi.org/10.1016/j.physe.2004.03.012
15. C.-H. Kuo, Y.-C. Yang, S. Gwo, M.H. Huang, Facet-dependent and Au nanocrystal-enhanced electrical and photocatalytic properties of Au–Cu_2O core–shell heterostructures. J. Am. Chem. Soc. **133**(4), 1052–1057 (2011). https://doi.org/10.1021/ja109182y
16. D.A. Engelbrecht, R. Synowicki, T. Tiedje, Luminescent coupling and efficiency of bifacial GaAs/Si tandem solar cells. Sol. Energy Mater. Sol. Cells **245**, 111800 (2022). https://doi.org/10.1016/j.solmat.2022.111800
17. J. Chen, H. Furukawa, Rapid and non-invasive detection of high-thickness glucose solution concentrations using quantum cascade laser-based transmission infrared spectroscopy. Infrared Phys. Technol. **131**, 104717 (2023). https://doi.org/10.1016/j.infrared.2023.104717
18. J. Ajayan et al., GaAs metamorphic high electron mobility transistors for future deep space-biomedical-military and communication system applications: a review. Microelectron. J. **92**, 104604 (2019). https://doi.org/10.1016/j.mejo.2019.104604
19. W.C. Chan, S. Nie, Quantum dot bioconjugates for ultrasensitive nonisotopic detection. Science **281**(5385), 2016–2018 (1998). https://doi.org/10.1126/science.281.5385.2016
20. M. Khatami, H.Q. Alijani, I. Sharifi, Biosynthesis of bimetallic and core–shell nanoparticles: their biomedical applications—a review. IET Nanobiotechnol. **12**(7), 879–887 (2018). https://doi.org/10.1049/iet-nbt.2017.0308
21. I.L. Medintz, H.T. Uyeda, E.R. Goldman, H. Mattoussi, Quantum dot bioconjugates for imaging, labelling and sensing. Nat. Mater. **4**(6), 435–446 (2005). https://doi.org/10.1038/nmat1390
22. A.J. Bennett et al., Giant Stark effect in the emission of single semiconductor quantum dots. Appl. Phys. Lett. **97**(3), 031104 (2010). https://doi.org/10.1063/1.3460912
23. E. Iqraoun, A. Sali, K. El-Bakkari, M.E. Mora-Ramos, C.A. Duque, Binding energy, polarizability, and diamagnetic response of shallow donor impurity in zinc blende GaN quantum dots. Micro Nanostruct. **163**, 107142 (2022). https://doi.org/10.1016/j.spmi.2021.107142
24. K. El-Bakkari, Polaron and conduction band non-parabolicity effects on the binding energy, diamagnetic susceptibility and polarizability of an impurity in quantum rings. Superlattices Microstruct. **148** (2020). https://doi.org/10.1016/j.spmi.2020.106729
25. I. Maouhoubi, R. En-nadir, K. El Bekkari, I. Zorkani, A. Ouazzani Tayebi Hassani, A. Jorio, Effects of applied magnetic field and pressure on the diamagnetic susceptibility and binding energy of donor impurity in GaAs quantum dot considering the non-parabolicity model's influence. Philos. Mag. 1–18 (2022). https://doi.org/10.1080/14786435.2022.2141469

26. J. El Khamkhami, E. Feddi, E. Assaid, F. Dujardin, B. Stébé, J. Diouri, Binding energy of excitons in inhomogeneous quantum dots under uniform electric field. Phys. E **15**(2), 99–106 (2002)
27. L. Belamkadem et al., The intensity and direction of the electric field effects on off-center shallow-donor impurity binding energy in wedge-shaped cylindrical quantum dots. Thin Solid Films **757**, 139396 (2022)
28. F. Dujardin, A. Oukerroum, E. Feddi, J. Bosch Bailach, J. Martínez-Pastor, M. Zazi, Effect of a lateral electric field on an off-center single dopant confined in a thin quantum disk. J. Appl. Phys. **111**(3), 034317 (2012)
29. G. Rezaei, N.A. Doostimotlagh, External electric field, hydrostatic pressure and conduction band non-parabolicity effects on the binding energy and the diamagnetic susceptibility of a hydrogenic impurity quantum dot. Phys. E **44**(4), 833–838 (2012). https://doi.org/10.1016/j.physe.2011.12.011
30. I. Maouhoubi, R. En-nadir, S. Maouhoubi, I. Zorkani, A.O. Tayebi Hassani, A. Jorio, The effects of lateral electric field, effective-mass position-dependent, and dimension on the donor-impurity binding energy and electron-impurity distance in GaAs-based quantum disk. Phys. E **148**, 115630 (2023). https://doi.org/10.1016/j.physe.2022.115630

Performance of III-N-Based Quantum Well Intermediate Band Solar Cells

Hassan Abboudi and Ilyass Ez-Zejjari

1 Introduction

Photovoltaic is one of the most promising alternative solutions for the production of green electricity, given its enormous potential on a global scale. Indeed, the photovoltaic technology has become a gigantic escort since researchers multiply their studies in order to obtain very powerful solar cells of very good quality, and this, by seeking the most suitable materials while minimizing their manufacturing cost. Since the fundamental investigations of Luque and Marti [1], the idea of a new generation of solar cells has grown tremendously. Obtaining above the Shockley–Queisser [2] limit by enhancing photovoltaic conversion efficiency more than 30% continues to be an issue for both academia and technology. To begin, let us provide a brief review of the key works that have been created in this topic. (For further details, please see Refs. [3–7].) The third generation corresponding to quantum well Intermediate Band Solar Cell (QW-IBSC) operates on the concept that photons with energy smaller than the band gap energy can supply to the photoelectric process of the material via a two-step process implying the intermediate band, i.e., valence band (VB) to intermediate band (IB) to conduction band (CB), boosting short-circuit current while maintaining open-circuit voltage. Additionally, III-Nitride materials have shown great potential for applications in electronics, optoelectronics, photovoltaics, and more. In particular, the (In,Ga)N material exhibits highly interesting physical properties. Its direct and tunable band gap offers a unique opportunity to develop high-efficiency solar cells. The band gap of InGaN can be tuned from 0.7 to 3.4 eV covering a broad portion of the solar spectrum, from the near-infrared to the

H. Abboudi (✉)
LPS, Sidi Mohamed Ben Abdellah University, Fez, Morocco
e-mail: hassan.abboudi@usmba.ac.ma

I. Ez-Zejjari
ENSAM, University Hassan II, Casablanca, Morocco
e-mail: lyass.ezzejjari-etu@etu.univh2c.ma

R. En-nadir (ed.), *Quantum Systems at the Nanoscale*, Lecture Notes in Nanoscale Science and Technology 45, https://doi.org/10.1007/978-3-032-06645-9_8

near-ultraviolet regions. InGaN alloys possess unique properties such as a tunable band gap, high carrier mobility, and strong radiation resistance [8, 9], which provide significant advantages in the design and fabrication of high-efficiency photovoltaic devices. Furthermore, the high absorption coefficient of InGaN (~ 10^5 cm^{-1}) [10, 11] also makes it particularly suitable for use in solar cells.

In this chapter, we focus on the study of intermediate band solar cells (IBSCs) incorporating quantum wells based on the (In,Ga)N material system. This type of structure has attracted increasing attention due to its potential to surpass the Shockley–Queisser efficiency limit, by introducing an intermediate band that allows the absorption of lower-energy photons.

The aim of this study is to thoroughly investigate the influence of various physical and technological parameters on the performance of these solar cells. The parameters considered include the alloy composition of (In,Ga)N, the geometrical dimensions of the quantum wells (well thickness, barrier width, periodicity), impurity concentrations, the nature of the confinement potential (abrupt or graded), as well as external factors such as temperature, applied electric field, hydrostatic pressure, and the intrinsic polarization field characteristic of III-nitride materials. The impact of these parameters will be evaluated in terms of key photovoltaic performance metrics, namely the open-circuit voltage (V_{oc}), short-circuit current density (j_{sc}), fill factor (FF), and the photovoltaic conversion efficiency (η). A detailed understanding of these effects is essential for optimizing quantum well-based devices and advancing the development of high-efficiency solar cells using III-nitride semiconductors.

2 Dual Confinement Effects in InGaN/GaN Quantum Well Solar Cells: Composition, Impurities, Structure, and Electric Field Effects

Let us begin our study by investigating the influence of structural parameters including active region dimensions, electric field intensity, In-composition, impurity position, and potential profiles on the photovoltaic characteristics of a p-GaN/i-(In,Ga)N/n-GaN (p-QW-n) structure. The finite element method (FEM) was used to numerically solve the Schrödinger equation. We found that particle and sub-gap energy levels are susceptible to well width, electric field, and impurity position. Potential profile shapes, such as rectangular (RQW) and parabolic (PQW), also play a significant role, with PQW profiles providing stronger particle confinement.

2.1 Theoretical Background

Our reference system is a p-GaN/(In,Ga)N/GaN-n solar cell illustrated in Fig. 1. Such a structure is characterized by GaN type layer with a thickness *L* (barrier with) and an intrinsic (In,Ga)N type layer with a thickness l (well width). Without losing generality, the influences of n(p)-type doping via the Poisson-equation are neglected in this paper. This structure constitutes a quantum-well intermediate-band solar cell (QW-IBSC) model, as depicted in Fig. 1. It allows the absorption of three photons via two steps, leading to much higher absorption than a conventional p-n structure. As shown in Fig. 1, E_C, E_v, E_{FC}, E_{Fv} and E_{FI} represent respectively the energies of the conduction band, valence band, and their quasi-Fermi levels, while E_{FI} Represents the IB quasi-Fermi level energy. CB and VB denote the conduction and valence bands, respectively. μ_{CI} and μ_{IV} are respectively the chemical potentials between the CB and the IB and μ_{CV} is the chemical potential between the CB and VB. E_{13}, E_{23} and E_{12} represent the bottom and top sub-bandgap energy, respectively.

The energy levels and their corresponding eigenfunctions in such a system are determined by solving the time-independent Schrodinger equation:

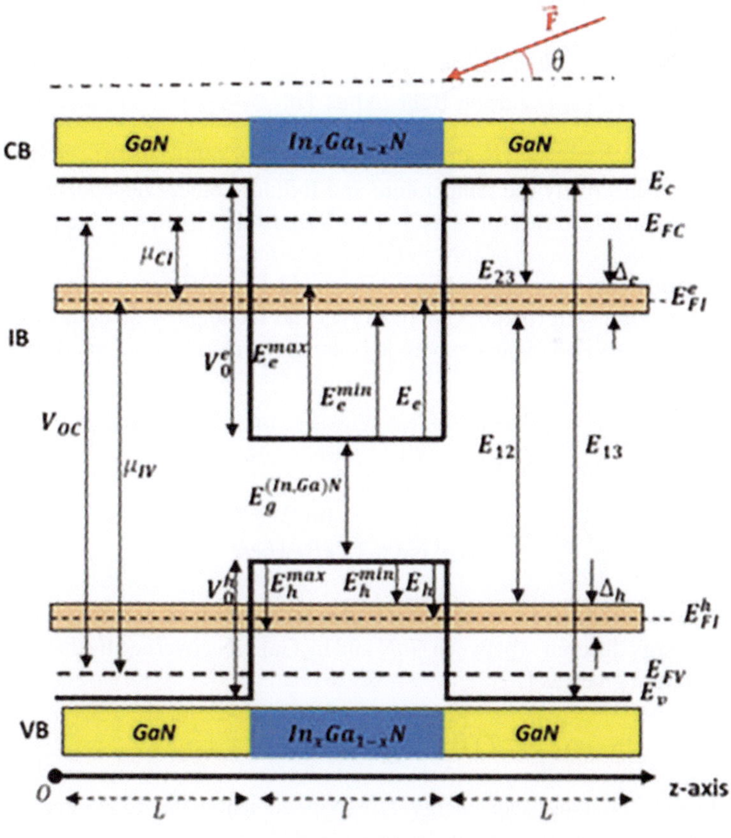

Fig. 1 Schematic diagram of GaN/InGaN/GaN QW one-intermediate band integrated in the conventional solar cell under study considering the quantum well width l, barriers height $V_o^{e,h}$, and built-in electric field $\vec{F}$ ($\theta = 0°$)

$$H\psi = E\psi \tag{1}$$

Within the parabolic band and the effective mass approximations, the Hamiltonian of a single particle in the $In_xGa_{1-x}N/GaN$ QW structure under the impurity and the electric field effects with regard to the indium content (x) and temperature (T) impacts is written as follows:

$$H_i = -\frac{\hbar^2}{2}\vec{\nabla}\left(\frac{1}{2m_i^*(x,T)}\vec{\nabla}\right) + V_0^i \pm \frac{e^2}{\varepsilon^*(x,T)\varepsilon_0|\vec{r}_i - \vec{r}_0|} \pm e\,F\,r_i\cos(\theta)\ (i \equiv e,h) \tag{2}$$

where $\overrightarrow{F}$ is the built-in electric field's intensity, θ is the angle between $\vec{F}$ and $\vec{r}_i$, e is the absolute electron charge, $\vec{r}_0(x_0, y_0, z_0)$ and $\vec{r}_i(x_i, y_i, z_i)$ are respectively the impurity and the particle position. m_i^* is the effective mass, ε_0 is the vacuum permittivity, ε^* is the relative dielectric constant while V_0^i is the particle-related related potential confinement.

We have adopted a dimensionless parameter considering the electric field effect to simplify. $\left(\mu = \frac{ea^*F}{R_b^*}\right)$ and the Bohr radius $\left(a^* = \frac{\varepsilon_0\hbar^2}{m^*\cdot e^2}\right)$ as the unit of the size and the effective Rydberg $\left(\mathrm{R}^* = \frac{m_b^* e^4}{2(4\pi\varepsilon_b^*\hbar)^2}\right)$ is used as unit energy. $m_i^*(x,T)$ and $\varepsilon^*(x,T)$ are respectively the temperature and Indium-content dependent effective-masses and main dielectric constants. For the ternary (In,Ga)N, they are given as the linear combination of those of InN and GaN materials.

In this study, we have also taken into account two different potential profiles. Parabolic quantum well (parab. QW or PQW) and rectangular quantum well (rect. QW or RQW). Their analytical expressions are given respectively as follows:

$$V_{0P}^i(z) = \begin{cases} \frac{4Q^i\Delta E_g(x,T)}{l^2}\left(z_i - L - \frac{l}{2}\right)^2 & \text{for } L \le z_i \le L + l \\ Q^i\Delta E_g(x,T) & elsewhere \end{cases} \tag{3}$$

$$V_{0R}^i(z) = \begin{cases} 0 & \text{for } L \le z \le L + l \\ Q^i\Delta E_g(x,T) & elsewhere \end{cases} \tag{4}$$

where $Q^i = 0.7\ (0.3)$ is the conduction (valence) band offset and $\Delta E_g(x,T)$ is the band-gap energy difference between GaN and $\mathrm{In}_x\mathrm{Ga}_{1-x}\mathrm{N}$ governed by the In-fraction and temperature given as [12]:

$$\Delta E_g(x,T) = E_g^{GaN}(T) - E_g^{\mathrm{In_xGa_{1-x}N}}(x,T) \tag{5}$$

The $\mathrm{In_xGa_{1-x}N}$ band gap energy is expressed using the following quadratic function of the In-concentration as follows [13, 14]:

$$E_g^{\mathrm{In_xGa_{1-x}N}}(x,T) = x\cdot E_g^{\mathrm{In}N}(T) + (1-x)\cdot E_g^{GaN}(T) - Dx(1-x) \tag{6}$$

Table 1 The main physical parameters used in our calculation [16]

Parameters	GaN	InN
E_g (eV)	3.42	0.78
ε*	8.68	11.6
α (meV/K)	0.593	0.245
β (K)	600	624
m^*_e/m_0	0.193	0.054
m^*_h/m_0	0.810	0.835
N^0_v (m^{-3})	8×10^{21}	10^{22}
N^0_c (m^{-3})	2.3×10^{20}	1.76×10^{20}

where D (= 1.43 eV) is the band gap bowing parameter. E_g^{InN} and E_g^{GaN} are respectively the band gap energies of InN and GaN. According to the Varshni formula, the temperature dependence of the band gap energy can be expressed as follows [15] (Table 1):

$$E_g(T) = E_g(0) - \frac{\alpha T^2}{T + \beta} \tag{7}$$

The presence of impurity makes the Schrödinger equation unsolvable analytically. Thus, the Finite Elements Method (FEM) is devoted to solving this problem considering a mono-directional mesh (calculation grid) composed of $3N + 1$ points with $N = 200$ points. It provides accurate results for the ground state and low-lying excited states of simple quantum systems such as the harmonic oscillator or the particle in a box. However, for more complex systems or higher excited states, the accuracy of the FEM solution may decrease due to the need for a finer mesh and higher-order basis functions. Overall, the accuracy of this method depends on the problem's complexity, the choice of numerical parameters, and the computational resources available.

2.2 Photonic Characteristics

To optimize photoelectric conversion efficiency a large absorption of the light spectrum is required. Indeed, three distinct optical transitions appear in such a structure: VB to IB, IB to CB and VB to CB. For the standard SC, only photons with energy higher than the band gap energy can generate electron–hole pairs. However, for IBSC system, the photons with energies less the band gap energy can be absorbed via the first and second optical transitions [17]. In light of what has been described above and for proper operation of IBSC. Certain assumptions must be maintained such as: the IB strip must be electrically isolated from external contacts, no current is extracted from the IB. All transitions between VB, IB and CB must be radiative, the quasi-Fermi levels corresponding to each band are constant and the device is thick enough to ensure a total absorption. Indeed, the photocurrent density, open circuit voltage and efficiency are the most interesting parameters of solar cells. Furthermore, under the condition of full-concentration sun light, the number of the absorbed and emitted photons determines the photo-generated density by the solar cell. Referring to the IBSC energy band diagram described in Fig. 1, j_{sc} can be expressed as follows [18]:

$$\begin{aligned}\frac{j_{sc}}{e} &= [N(E_{13}, \infty, T_s, 0) - N(E_{13}, \infty, T_c, \mu_{CV})] \\ &\quad + [N(E_{23}, E_{12}, T_s, 0) - N(E_{23}, E_{12}, T_c, \mu_{CI})]\end{aligned} \tag{8}$$

where T_S and T_C are respectively the surface temperatures of the sun and the solar cell, μ_{CV} is the chemical potential between the CB and the VB, μ_{CI} is the chemical potential between the IB and CB. Notice that all transition energies (E_{13}, E_{12} and E_{23}) are specified in Fig. 1. Moreover, according to the Roosbroeck–Shockley formula, the flux N of photons leaving an object at temperature T, is be expressed as [19]:

$$N(E_a, E_b, T, \mu) = \frac{2\pi}{h^3c^2} \int\limits_{E_a}^{E_b} \frac{E^2 dE}{e^{\frac{(E-\mu)}{k_B T}} - 1} \tag{9}$$

where E_a and E_b are respectively, the lower and upper energy limits of the photons flux for the corresponding transitions, T is the temperature, h is the Plank's constant and c is the light speed in vacuum, $k_B \left(\approx 1.38 \times 10^{-23}\,\text{J/K}\right)$ is the Boltzmann's constant, and μ is the chemical potential of the transition.

On the other hand, the *IBSC* output voltage (V_{oc}) of a $p - i - n$ solar cell can be written as follows [20]:

$$V_{oc} = \mu_{CV} = \mu_{CI} + \mu_{IV} \tag{10}$$

To be more realistic, mainly in previous studies related to the theoretical efficiency of solar cells, the fill factor was often assumed to be 1, which means that the entire surface of the solar cell was assumed to be active and capable of converting sunlight into electric energy. However, in reality, the fill factor is always less than 1 because there are always areas on the cell surface that are inactive; however, we have used a fill factor (FF) expression depending on the open-circuit voltage as follows with $V_{th} = \frac{K_B T}{q}$ [21]:

$$FF = \frac{\frac{V_{oc}}{V_{th}} - \ln\left(\frac{V_{oc}}{V_{th}} + 0.72\right)}{1 + \frac{V_{oc}}{V_{th}}} \tag{11}$$

Therefore, the photovoltaic conversion efficiency can be obtained using the output voltage and photo-current density as follows [22]:

$$\eta = \frac{V_{oc} \cdot J_{sc} \cdot FF}{P_{in}} \tag{12}$$

We notice that P_{in} is the incident sun power per area unit given by the Stefan–Boltzmann's law $P_{in} = \sigma T_s^4$, where $\sigma = 5.67 \times 10^{-8}\,\mathrm{Wm^{-2}\,K^{-4}}$.

3 Results and Discussion

Figure 2 presents the variations of the open-circuit voltage V_{oc} as a function of the internal electric field and indium In composition at room temperature for both rectangular and parabolic quantum well profiles. It is observed that V_{oc} increases linearly with the electric field and decreases linearly with increasing In-content, regardless of the quantum well profile or impurity position. Notably, the right panel of Fig. 2 shows that the parabolic profile, combined with a high electric field, yields the highest V_{oc}. Conversely, the left panel indicates that the rate of decrease in V_{oc} with increasing In-content is more pronounced than the rate of increase with the electric field.

Fig. 3 illustrates the variation of the photo-generated current densityj_{sc} as a function of the electric field and In composition for both rectangular and parabolic profiles at room temperature, with L = 3l = 3. The results show thatj_{sc}decreases as these

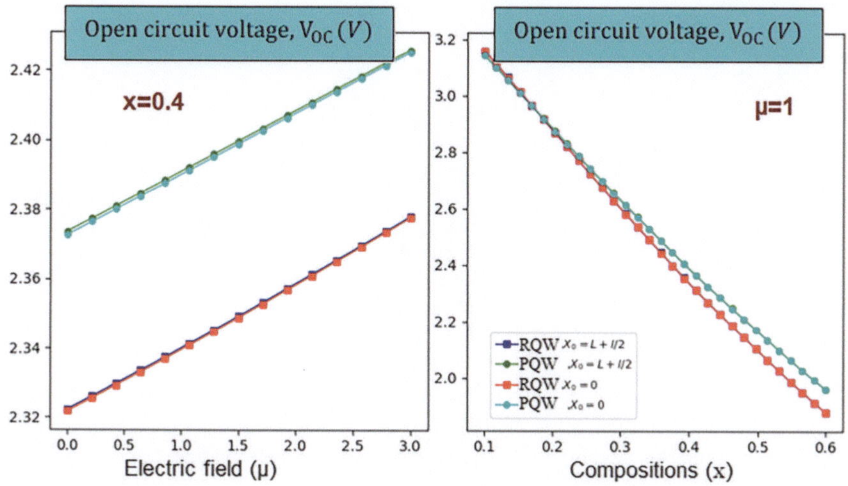

Fig. 2 The open circuit voltage variation versus the electric field intensity (left) and In-content (right) for two different impurity positions and two different profiles. T = 300 K, l = 1 and L = 3 for $\theta = 0°$

parameters increase, independent of the profile shape or impurity position, while the parabolic profile systematically exhibits a higherj_{sc}than the rectangular one.For example, as the electric field parameter μ increases from 0 to 3, the short-circuit current density decreases from 74.5 to 53.14 mA/cm^2 for the parabolic profile, and from 60.92 to 38.66 mA/cm^2 for the rectangular profile. This corresponds to a reduction of approximately 28.7% and 36.5%, respectively, indicating a more pronounced sensitivity of the rectangular profile to electric field variations. Similarly, with increasing In-composition from 10 to 60%, j_{sc} decreases from 90 to 28 mA/cm^2 for the rectangular structure and from 92 to 52 mA/cm^2 for the parabolic structure. These results indicate a substantial reduction of approximately 69% and 43%, respectively, highlighting that the rectangular configuration is more sensitive to In-content variations than the parabolic one. Nevertheless, both profiles exhibit a similar declining trend in

j_{sc} with increasing electric field, independent of impurity loc.

After analysing the open-circuit voltage and photo-generated current density, we now focus on the most critical parameter for quantum well intermediate band solar cells (QW-IBSCs): the photovoltaic conversion efficiency. Figure 4 presents the simulated efficiency as a function of In-composition (right panel) and electric field strength (left panel), computed at room temperature for a structure with dimensions L = 3 and l = 3. It is evident that the efficiency trends closely follow those of V_{oc} and J_{sc}, underscoring their dominant influence. The highest efficiency is achieved with the parabolic quantum well profile, reflecting its superior carrier extraction and confinement characteristics. Moreover, as the In-content increases, the efficiency decline is more pronounced for the rectangular profile than for the parabolic one, indicating greater sensitivity to compositional changes. Importantly, the maximum photovoltaic

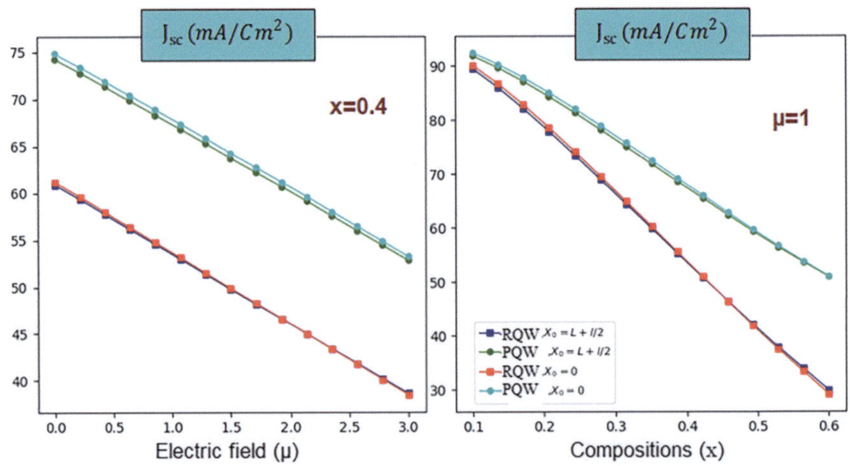

Fig. 3 The variation of the short-current density as a function of electric field strength (left) and In-content (right) for two different impurity position considering the shape effect at room temperature. T = 300 K, l = 1 and L = 3 for $\theta = 0°$

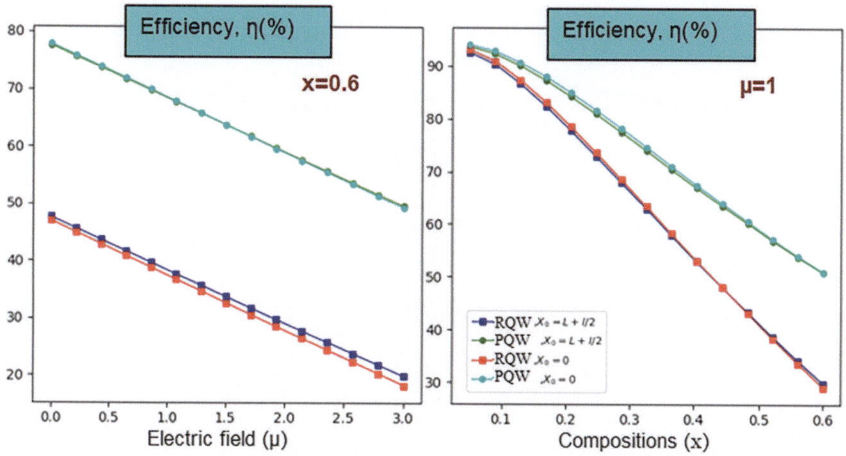

Fig. 4 Photovoltaic conversion efficiency versus the electric field (left) and In-content (right) at room temperature considering two different impurity positions and two types of potentials. $l = 1$ and $L = 3$ for $\theta = 0°$

efficiency is consistently obtained in the absence of an applied electric field ($\mu = 0$), regardless of other structural parameters.

As an illustrative example, when the electric field strength is increased from $\mu = 0$ to $\mu = 3$, the photovoltaic efficiency for the parabolic potential profile decreases from 77.57 to 49.43%, representing a reduction of approximately 36.27%. In the case of an impurity positioned at the left barrier edge (on-center), the efficiency for the rectangular profile drops from 47.87% (47.17%) to 19.90% (17.99%), corresponding to a decline of 58.43% (61.86%). The left panel of Fig. 4 illustrates the efficiency variations as a function of In-content, under a fixed electric field intensity, while accounting for the effects of both impurity position and potential profile shape. A consistent decrease in efficiency is observed with increasing In-content, independent of impurity location. Specifically, as the In-composition rises from 10% to 60%, the efficiency for the parabolic profile drops from 91.99% to 50.42%, and from 89.78% to 29.08% for the rectangular profile amounting to reductions of approximately 45.19% and 67.61%, respectively. These efficiency losses can be attributed to the shrinking of the inter-band transition energy, which shifts toward lower energies with increasing In-content. This shift modifies both the position and width of the intermediate band (IB), thereby adversely impacting carrier dynamics and overall device performance.

The orientation of the built-in electric field relative to the growth direction specifically its component along the z-axis plays a critical role in determining the optical and electronic behavior of nanostructures such as quantum wells. Figure 5 illustrates the effect of the electric field angle θ on the photovoltaic efficiency of the device, with all other parameters held constant. Both rectangular and parabolic profiles exhibit a comparable efficiency response to changes in field orientation. Notably, the efficiency remains unchanged at $\mu = 0$ (corresponding to $\theta = \pi/2$), and exhibits a nonlinear increase as the angle varies, eventually converging back to a maximum at $\theta = \pi/$

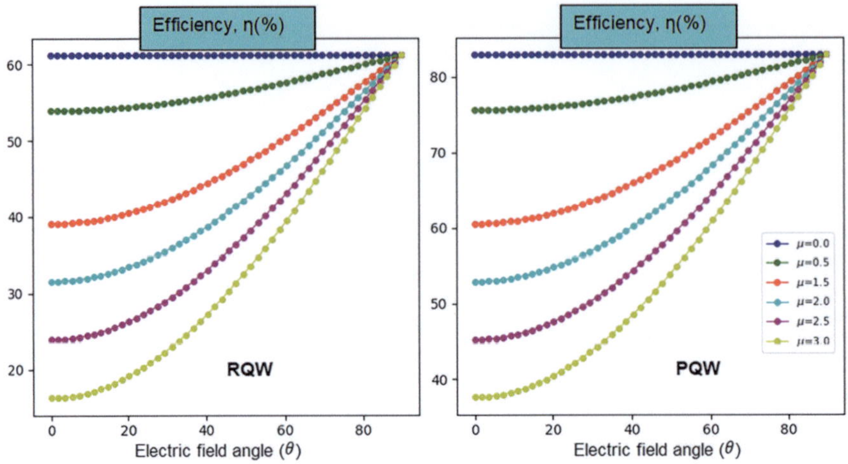

Fig. 5 On-center impurity-related photovoltaic conversion efficiency versus the angle θ for different values of electric field strength and two different forms of potentials at room temperature. $x = 0.5$, $l = 2$ and $L = 4$

2. This behavior aligns with the suppression of the quantum confined Stark effect, which is minimal when the electric field is perpendicular to the growth axis. Furthermore, the influence of the electric field orientation on efficiency is more pronounced at smaller angles, while it becomes less significant as the angle approaches π/2 [23].

Overall, variations in open-circuit voltage V_{oc}, short-circuit current density j_{sc}, and fill factor FF induced by changes in electric field orientation collectively influence the output power and, consequently, the overall efficiency of the solar cell. Thus, optimizing these parameters is essential for enhancing device performance. It is important to note that the present results, derived under idealized conditions, likely represent an upper performance bound compared to practical implementations. In real devices, factors such as non-radiative recombination processes, structural imperfections, and fabrication-induced stresses although not explicitly addressed in this study can significantly degrade the actual efficiency and must be considered in comprehensive performance assessments.

4 Investigation of Temperature-Dependent Photovoltaic Performance in (In,Ga)N Single-Intermediate Band Quantum Well Solar Cells for Different Shapes

4.1 Theory and Applied Model

Recent advancements in semiconductor nanostructure fabrication particularly through techniques such as molecular beam epitaxy (MBE) have enabled precise control over the growth of quantum wells (QWs) with diverse geometrical profiles, including rectangular (RQWs), parabolic (PQWs), and triangular (TQWs) configurations. These varied QW shapes exhibit distinct quantum confinement effects that are of considerable interest for applications in microelectronic and optoelectronic devices. Despite extensive theoretical and experimental efforts in this area, few studies have systematically explored the combined influence of quantum well shape, impurity states, and heavy-hole contributions on device performance. In this work, we present a comprehensive analysis of (In,Ga)N-based quantum well structures with different potential profiles, incorporating the often-neglected roles of impurity interactions and heavy-hole effects, while also examining their temperature-dependent photovoltaic behavior. To the best of our knowledge, this represents one of the first investigations to concurrently address these factors in the context of intermediate band solar cell design.

In order to numerically calculate the efficiency of the photovoltaic conversion, we should comply with the following postulates specified by Luque and Marti [24]:

- The solar cells are thick enough to ensure complete photon absorption.
- Non-radiative transitions are not allowed.
- Carrier mobility is assumed to be sufficiently high.

4.2 Results and Discussion

The key photovoltaic performance indicators namely the open-circuit voltage V_{oc}, short-circuit current density j_{sc}, fill factor FF, and overall power conversion efficiency (PCE) are analyzed in this section. Figure 6 illustrates the variation of V_{oc}, with temperature for the three quantum well potential profiles: rectangular (RQW), parabolic (PQW), and triangular (TQW). The simulation results reveal a clear trend of decreasing V_{oc}, with increasing temperature across all potential configurations. Among the profiles studied, the triangular quantum well exhibits the highest V_{oc}, values across the temperature range. Specifically, V_{oc} decreases from 2.02 V to 1.91 V for TQW, from 1.99 V to 1.88 V for PQW, and from 1.86 V to 1.75 V for RQW, corresponding to relative reductions of approximately 5.2%, 5.3%, and 5.6%, respectively, over the 100–500 K temperature interval. These findings indicate that while all quantum well shapes exhibit thermal degradation of the open-circuit voltage, the

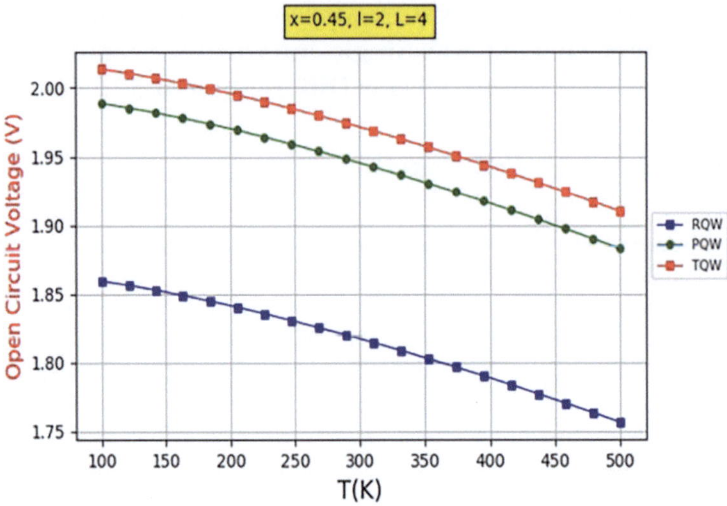

Fig. 6 The open-circuit voltage versus temperature for three different forms of potentials for on-center impurity GaN/$In_{0.45}Ga_{0.55}N$/GaN GaN QW-IBSC

triangular profile offers a comparatively enhanced thermal stability. Overall, the data suggest that temperature-induced reductions in V_{oc}, are a critical factor to consider in the design of quantum well-based photovoltaic devices, irrespective of potential well geometry.

We now turn our attention to the photogenerated current density j_{sc}, a critical parameter that strongly influences the overall power conversion efficiency of photovoltaic devices. Figure 7 presents the temperature-dependent behavior of j_{sc}, for the three quantum well potential profiles: triangular (TQW), parabolic (PQW), and rectangular (RQW). The simulation results indicate a modest decrease in j_{sc}, with increasing temperature for all potential shapes. Among the profiles, TQW consistently yields the highest j_{sc}, values across the examined temperature range. At room temperature (300 K), the calculated current densities are 53.18 mA cm^{-2} for TQW, 39.38 mA cm^{-2} for PQW, and 32.14 mA cm^{-2} for RQW. It is noteworthy that the relative decrease in j_{sc}, with temperature is less pronounced than that observed for the open-circuit voltage V_{oc}, Specifically, the current density decreases by approximately 2.14%, 3.57%, and 4.37% for TQW, PQW, and RQW, respectively, over the temperature range from 100 K to 500 K. These results highlight the superior thermal robustness of TQW structures in preserving current generation under elevated temperatures, underscoring their potential advantage in high-temperature photovoltaic applications.

As previously discussed, our analysis departs from the idealized scenario frequently assumed in related studies, in which the fill factor FF is approximated as unity (FF = 1). Instead, we employ a more realistic model where the FF is temperature-dependent and governed by the open-circuit voltage, as described by Eq. (19). Figure 8 illustrates the variation of FF with temperature for the three

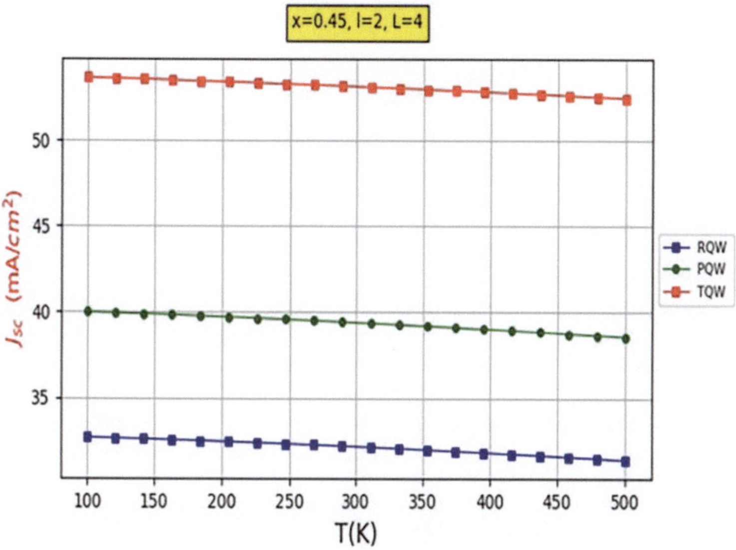

Fig. 7 The photo-generated current density versus the temperature for three profiles of potential for on-center impurity $GaN/In_{0.45}Ga_{0.55}N/GaN$ GaN QW-IBSC

quantum well potential profiles (TQW, PQW, and RQW). As anticipated, the fill factor exhibits a similar trend to that of V_{oc}, decreasing gradually with increasing temperature for all cases. Notably, the TQW configuration consistently achieves the highest FF across the studied temperature range, followed by PQW and RQW. At room temperature 300 K, the FF values are 93.12% for TQW, 93.05% for PQW, and 92.66% for RQW. Although the temperature-induced decline in FF is relatively small, it is observable across all profiles in the 100–500 K range. These results further confirm the enhanced performance stability of triangular quantum wells under thermal stress, reinforcing their suitability for high-efficiency photovoltaic applications.

Figure 9 presents the simulated photovoltaic conversion efficiency as a function of temperature for the three quantum well potential profiles: triangular (TQW), parabolic (PQW), and rectangular (RQW). The calculations were performed for a structure incorporating an on-center donor impurity and a quantum well width defined by $L = 2l = 2a_b^*$. The results clearly indicate that the photovoltaic efficiency is predominantly governed by the open-circuit voltage V_{oc}, and the photogenerated current density j_{sc}, as both parameters exhibit strong temperature dependence. Consistent with the behavior of V_{oc}, and j_{sc}, the efficiency decreases with increasing temperature for all potential profiles, highlighting the detrimental effect of thermal fluctuations on device performance. Among the profiles studied, the TQW configuration yields the highest conversion efficiency, followed by PQW and RQW. At room temperature, the efficiency values are approximately 71.64% for TQW, 52.46% for

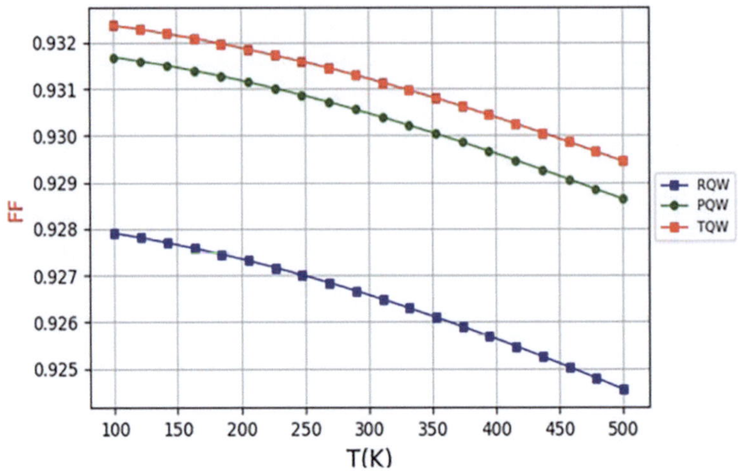

Fig. 8 The fill Factor as a function of the temperature for three potential profiles for on-center impurity $GaN/In_{0.45}Ga_{0.55}N/GaN$ QW-IBSC

PQW, and 39.74% for RQW. These values notably exceed the performance of conventional single-junction solar cells, demonstrating the potential advantages of incorporating nanostructured absorbers such as quantum wells into the intrinsic region to create an intermediate band (IB). This intermediate band facilitates the absorption of sub-bandgap photons and enhances carrier generation, thereby boosting overall efficiency.

The high efficiency observed in our simulations can be attributed to the near-complete absorption of the solar spectrum achieved through the optimal positioning and size of the intermediate band under the chosen structural parameters. As temperature increases from 100 to 500 K, the efficiency declines by approximately 9.4% for TQW, 9.2% for PQW, and 7.6% for RQW, confirming the general sensitivity of quantum well-based devices to thermal effects, though TQW maintains superior performance across the temperature range.

It is important to note that the simulations presented in this study were conducted under idealized conditions, assuming full light concentration and neglecting non-ideal effects. As a result, the predicted photovoltaic efficiencies are expected to exceed those achievable under standard experimental or operational conditions. Moreover, it should be emphasized that the influence of built-in electric fields such as those arising from band bending or polarization effects has not been considered in the present analysis. These fields can play a significant role in carrier separation and transport, and their omission may lead to deviations from real-world device performance. Therefore, while the results provide valuable insight into the theoretical potential of the proposed quantum well structures, they represent an upper performance limit rather than practical efficiency outcomes.

In summary, we have investigated a $GaN/In_{0.45}Ga_{0.55}N/GaN$ quantum well intermediate band solar cell (QW-IBSC) incorporating an on-center donor impurity, with

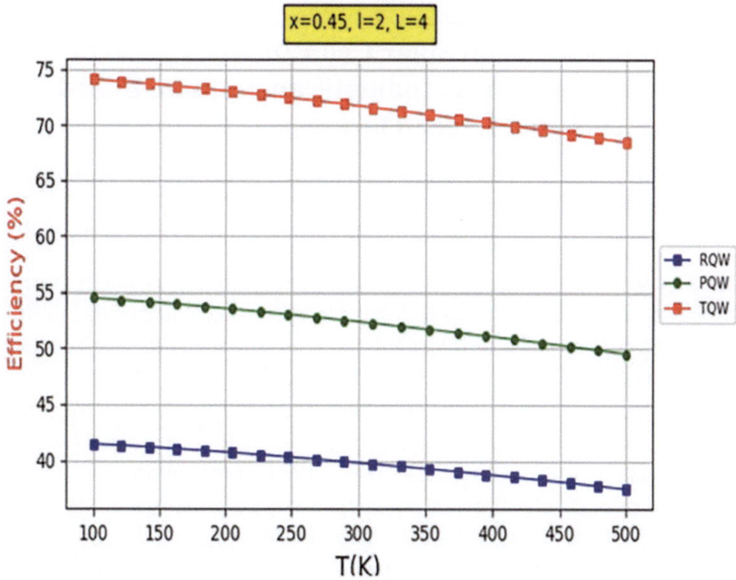

Fig. 9 The photovoltaic conversion efficiency versus the temperature and for three different forms of potentials for on-center impurity

the objective of enhancing photon absorption across a broad spectral range. The study focused on the effects of different quantum well confinement profiles rectangular, parabolic, and triangular as well as temperature dependence on the device's photovoltaic performance. The numerical simulations were conducted under ideal full-spectrum illumination conditions using the effective mass approximation and finite difference method, with optimized well and barrier widths and alloy compositions.

Our results demonstrate that the triangular potential profile yields the highest conversion efficiency, achieving 71.64% at room temperature, compared to 52.46% and 39.79% for the parabolic and rectangular profiles, respectively. Furthermore, while the photovoltaic parameters deteriorate with increasing temperature for all geometries, the extent of degradation is notably shape-dependent. Specifically, the triangular configuration exhibits a 9.4% drop in efficiency from 100 K to 500 K, compared to 9.2% and 7.6% for the parabolic and rectangular cases, respectively.

Overall, this study provides a simplified yet insightful model that highlights the critical influence of quantum well geometry and thermal effects on the performance of nanostructured IBSCs. These findings offer valuable guidance for the design and optimization of high-efficiency solar cells employing quantum-confined intermediate band structures.

5 Comprehensive Optimization Strategies for Intermediate-Band Solar Cell Efficiency: Effects of Hydrostatic Pressure, Indium Composition, Built-In Electric Fields, and Layer Thickness

5.1 *Theory and Models*

In this section, we examine the effects of hydrostatic pressure, impurity positioning, dimensional variations, and compositional changes—particularly the indium (In) content—on the performance of intermediate band solar cells (IBSCs) employing a wurtzite (WZ) Ga-face GaN/InN/(In,Ga)N/GaN strained semi-graded quantum well (SGQW) architecture. The heterostructure is epitaxially grown along the [0001] crystallographic direction (c-axis) of the bulk material. To accurately capture the physical behavior of the system, both spontaneous polarization and piezoelectric effects are incorporated into the analysis, acknowledging the inherently polar characteristics of III-nitride semiconductors. The SGQW configuration consists of two coupled quantum wells (QWs): the first, designated as the left quantum well (LQW), is composed of InN with a thickness (l_1), while the second, the right quantum well (RQW), is formed from InGaN and has a thickness (l_2). These QWs are symmetrically confined by GaN barrier layers of thickness L, and the entire active region is embedded within the intrinsic region (i-region) of the device structure. The c-axis of the wurtzite lattice is oriented parallel to the growth axis (z-axis). Figure 10 presents a detailed schematic illustration of the GaN/InN/InGaN/GaN-based IBSC, highlighting the active region, electrical contacts, and substrate configuration.

The energy levels and their corresponding wave functions are determined through the solution of the one-dimensional time-independent Schrödinger equation, expressed as follows:

$$H\psi(z) = E\psi(z) \tag{13}$$

The Hamiltonian governing the behavior of a single particle (i.e., electron, hole) within the studied system is defined as presented in Eq. (14). This equation considers factors such as a mobile hydrogen-like impurity, composition, pressure, and the built-in field (BEF) effect. This equation is calculated within the framework of both the one-band parabolic theory and the effective mass approach. It has been numerically solved using the Finite Element Method (FEM) due to increased complexity arising from the inclusion of the Coulombian term (impurity), rendering it almost analytically unsolvable [25–27].

$$\mathrm{H} = -\frac{\hbar^2}{2}\vec{\nabla}\left(\frac{1}{m_i^*(x,p,z)}\vec{\nabla}\right)\psi_i(z) - \frac{\alpha_i e^2}{\varepsilon^*(x,p,z)\varepsilon_0|\vec{r}_i-\vec{r}_0|}\psi_i(z) + V_T(x,p,z)\psi_i(z) = E_i\psi_i(z)$$
$$(i = e, h) \tag{14}$$

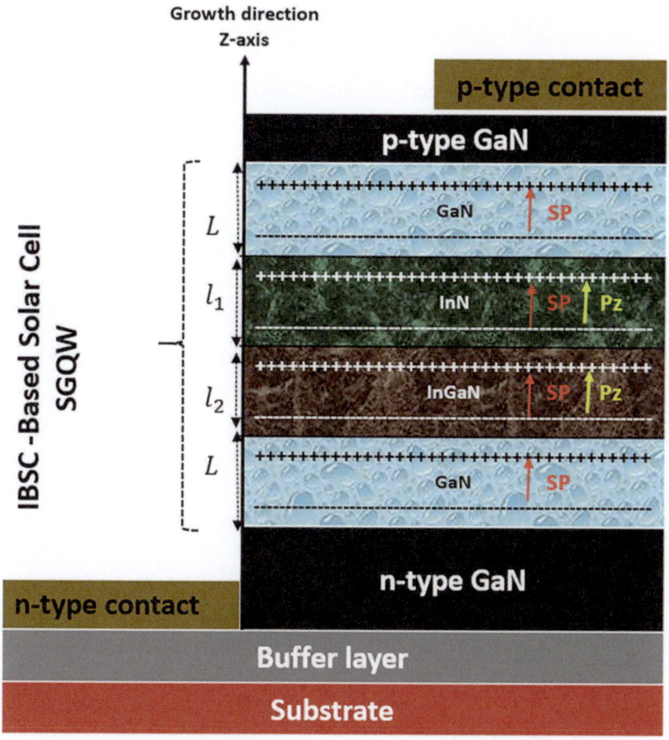

Fig. 10 Illustrates a comprehensive schematic of the GaN/InN/InGaN/GaN-based intermediate band solar cell (IBSC)

where e is the electron charge and $\alpha_{e(h)} = 1(-1)$, $|\vec{r}_i - \vec{r}_0|$ denotes the electron-impurity distance while ε_0 represents the dielectric constant of the vacuum. $m_i^*(x, p, z)$ and $\varepsilon^*(x, p, z)$ are the electron effective mass and relative dielectric constant, both contingent upon the pressure, composition, and displacement of the particle.

$V_T(x, p, z)$ is the pressure and position-dependent total potential energy expressed as follows:

$$V_T(x, p, z) = V_i(x, p, z) + \Phi(x, p, z) \tag{15}$$

The first term $V_i(x, p, z)$ is the electron (hole) confinement potential due to the band offset in WZ GaN/InN/InGaN/GaN SGQW expressed as in Ref. [28]:

$$V_i(x, p, z) = Q_i \Delta E_{g\,j}^{\,\Gamma} \tag{16}$$

With,

$$\Delta E_{g\,j}^{\;\Gamma} = \Delta E_{g\,j}^{\;\Gamma}\,(\text{unstrained}) + \delta E_{c\,j}^{\;\text{strain}} \tag{17}$$

The band-offset ratio, defined as the ratio between the conduction band offset (ΔE_c) and the valence band offset (ΔE_v), is assumed to be $Q_i = 0.63\,(0.37)\,(i \equiv e, h)$.

Where $\Delta E_{g\,j}^{\;\Gamma}$ (*unstrained*) is the band-gap energy between the barrier and the well at Γ point. The second term $\Phi(x, p, z)$ is the static electric potential. Taking into account the lattice mismatch between LQW, RQW, and barrier, the static electric potential term, denoted as $\Phi(x, p, z)$, encompasses the influence of the internal electric field on the polarization charges [28].

$$\delta E_{c\,j}^{\;strain} = 2a_c\left(\frac{a^w}{a^b} - 1\right)\left(1 - \frac{c_{12}^w}{c_{11}^w}\right) \tag{18}$$

With,

$$\begin{cases} j = 1, \; For\; 0 \le z \le L \\ j = 2, \; For\; L \le z \le L + l_1 \\ j = 3, \; For\; L + l_1 \le z \le L + l_1 + l_2 \\ j = 4, \; For\; L + l_1 + l_2 \le z \le 2L + l_1 + l_2 \end{cases} \tag{19}$$

5.2 Built-In Electric Field

According to Eq. (3), the second term is given as follows:

$$\Phi(x, p, z) = \alpha e F_j(x, p, z) z \tag{20}$$

F_j represents the BEF in various regions of our structure attributable to piezoelectricity and spontaneous polarizations. The orientation-dependent F_j electric field arises from the interplay of piezoelectricity, spontaneous polarization, crystal polarity, and strains within the SGQW system. The alignment of both piezoelectricity and spontaneous polarization coincides with the z-direction. As a result, the effect of BEF aligns with the growth direction, as depicted in Fig. 1a [29].

For simplification, we will adopt the approach of Takeuchi et al. [30] and disregard the intricate strains arising in GaN layers due to the lattice and thermal mismatch between GaN and the substrate. Furthermore, we will not account for the alterations in the lattice constant of the thin strained $In_xGa_{1-x}N$ layer caused by biaxial compressive stress [31], we obtain:

$$2LF_{GaN} + l_1 F_{InN} + l_2 F_{InGaN} = 0 \tag{21}$$

Over this computation, our initial consideration involves a thin layer of InN wells that have been meticulously grown atop a substantial layer of GaN. Subsequently, a slender layer of InGaN wells is coherently developed atop a thick InN layer along the [0001] direction. This is followed by a GaN barrier layer. The strain within the InGaN layer varies according to distinct indium composition and diverse growth conditions. Moreover, it is crucial to note that the formula for the strength of the F_j due to the BEF being derived based on the assumption that the potential energies at the extreme left and right of our SGQW structure are equal. Furthermore, we expand the first-order polarization of the integrated electric field and apply the boundary conditions that ensure continuity of the electric displacement vector at the hetero-interfaces within our SGQW structure, we can obtain the following expression [32, 33]:

$$\begin{cases} \varepsilon_{InN}\varepsilon_0 F_{InN} = \varepsilon_{GaN}\varepsilon_0 F_{GaN} + P_{GaN} - P_{InN} \\ \varepsilon_w \varepsilon_0 F_w = \varepsilon_{InN}\varepsilon_0 F_{InN} + P_{InN} - P_w \end{cases} \tag{22}$$

The size-dependent behavior of the built-in electric field (BEF) for various regions within the studied system (SGDW) is formulated as follows:

$$F_j = \begin{cases} F_1 = F_{GaN} = \dfrac{(l_1\varepsilon_w P_{InN} + l_2\varepsilon_{InN} P_w) - P_{GaN}(l_1\varepsilon_w + l_2\varepsilon_{InN})}{\varepsilon_0[2L\varepsilon_{InN}\varepsilon_w + l_1\varepsilon_{GaN}\varepsilon_w + l_2\varepsilon_{GaN}\varepsilon_{InN}]} & 0 \le z \le L \\ F_2 = F_{InN} = \dfrac{(2L\varepsilon_w P_{GaN} + l_2\varepsilon_{GaN} P_w) - P_{InN}(2L\varepsilon_w + l_2\varepsilon_{GaN})}{\varepsilon_0[2L\varepsilon_{InN}\varepsilon_w + l_1\varepsilon_{GaN}\varepsilon_w + l_2\varepsilon_{GaN}\varepsilon_{InN}]} & L \le z \le L + l_1 \\ F_3 = F_w = \dfrac{(2L\varepsilon_w P_{GaN} + l_2\varepsilon_{GaN} P_{InN}) - P_w(2L\varepsilon_{InN} + l_1\varepsilon_{GaN})}{\varepsilon_0[2L\varepsilon_{InN}\varepsilon_w + l_1\varepsilon_{GaN}\varepsilon_w + l_2\varepsilon_{GaN}\varepsilon_{InN}]} & L + l_1 \le z \le L + l_1 + l_2 \\ F_4 = 0 & z \ge L + l_1 + l_2)) \end{cases} \tag{23}$$

where (L, ε_{GaN}), (l_1, ε_{InN}) and (l_2, ε_w) denote respectively the layer thickness and the dielectric constant of the LQW, RQW, and the barriers. Hence, the total polarization (P_ν), is given by the sum of both the piezo and spontaneous polarizations and it is expressed as follows:

$$P_\nu = P_\nu^{sp} + P_\nu^{pz} \tag{24}$$

where ν denotes either GaN or InN while w denotes InGaN alloy.

The second-order expression for the spontaneous polarization of random ternary group III-nitride alloys, denoted by x, is given in units of C/m^2 [34].

$$P_{In_xGa_{1-x}N}^{sp} = xP_{InN}^{sp} + (1-x)P_{GaN}^{sp} + 0.037x(1-x) \tag{25}$$

The piezoelectric polarization of the binary compounds is formulated as follows [35]:

$$\begin{cases} P_{InN}^{pz} = 1.373\varepsilon + 7.559\varepsilon^2 \\ P_{GaN}^{pz} = -0.918\varepsilon + 9.541\varepsilon^2 \end{cases} \tag{26}$$

The piezoelectric field polarization along the c-axis resulting from the mismatch between well and barrier materials is provided by the following equation:

$$P^{pz}_{InxGa_{1-x}N} = e_{31}\left(\varepsilon_{xx} + \varepsilon_{yy}\right) + e_{33}\varepsilon_{zz} \tag{27}$$

where e_{ij} and ε_{ij} are the piezoelectric constants and the strain elements of (In,Ga)N materials, respectively. It is expressed analytically as follows:

$$\varepsilon_{xx} = \varepsilon_{yy} = \varepsilon = \tfrac{a_b - a_w(x)}{a_w(x)} \tag{28}$$

$$\varepsilon_{zz} = -2\tfrac{c_{13}}{c_{33}}\varepsilon_{xx} \tag{29}$$

a_b and a_w are the lattice parameters of the wells and the barriers, respectively. C_{13} and C_{33} are the elastic constants. Hence, the revised expression for the piezoelectric polarization is presented as follows:

$$P^{pz}_{InxGa_{1-x}N} = 2\left(\frac{a^b - a^w}{a^w}\right)\left(e_{31} - e_{33}\frac{c_{13}}{c_{33}}\right) \tag{30}$$

5.3 Results and Discussion

Figure 11 presents the dependence of the open-circuit voltage (V_{oc}) on the indium (In) composition at a temperature of T = 300 K, considering the on-center impurity configuration and adopting an initial localization length $l_1(0) = 2.0a_b^*$. Panel (11a) specifically illustrates the variation of V_{oc} for four different barrier widths, with fixed pressure P = 10 GPa and $l_2(0) = 1.0a_b^*$. The results demonstrate a non-linear reduction in V_{oc} as the In fraction increases. This reduction is strongly influenced by the enhanced spontaneous and piezoelectric polarization effects, which intensify with higher In content.

Moreover, the data reveal a slight decline in V_{oc} as the barrier width increases. Specifically, V_{oc} decreases from 1.28 to 1.23 V—a relative drop of approximately 3.9%—as the barrier width L(0) increases from 2.0 to 8.0 nm. This behavior underscores the limited influence of barrier width on V_{oc}, which is primarily governed by the energy difference between the conduction band (CB) and the valence band (VB), corresponding to the intrinsic band gap of the host material. Consequently, variations in L(0) induce only marginal changes in V_{oc}.

In panel (11b), the variation of the open-circuit voltage V_{oc} is illustrated under constant conditions: P = 10 GPa and L(0) = 4.0a_b^*. The analysis focuses on four distinct values of the InGaN layer width, denoted as $l_2(0)$. The results clearly show a decreasing trend in V_{oc} with increasing indium composition. This reduction is primarily attributed to the elevated defect density induced by higher In content, in

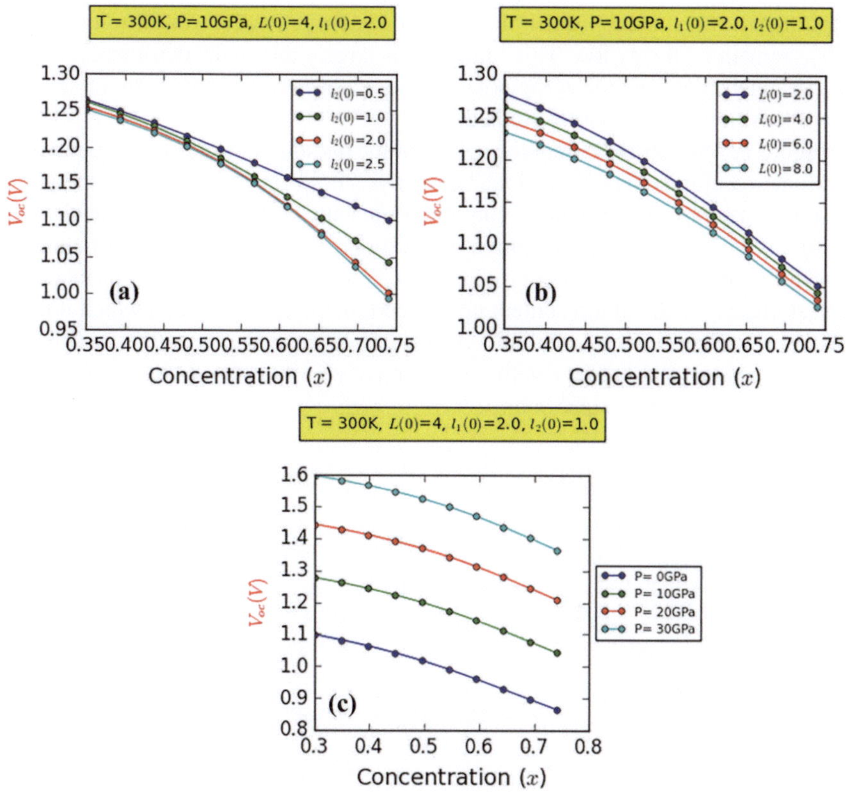

Fig. 11 Variation of the open circuit current (V_{oc}) as a function of the indium-concentration (x) for different values of barrier's width (**a**), InGaN-layer's width (**b**), and hydrostatic pressure (**c**)

addition to the pronounced effects of spontaneous and piezoelectric polarization. Moreover, the figure reveals the presence of a critical indium composition, xc = 52.34%, which marks a transition in behavior. For x < xc, V_{oc} exhibits an approximately linear decrease, irrespective of the InGaN layer width. Conversely, for x > xc, the width of the InGaN layer significantly influences the V_{oc} response. Except for the case of $l_2(0) = 0.5$ nm, the voltage decline becomes non-linear. Specifically, as the In fraction increases from 0.35 to 0.74, the decrease in V_{oc} is approximately 13.07% for $l_2(0) = 0.5a_b^*$, and 20.63% for $l_2(0) = 2.5a_b^*$.

In panel (11c), the variation of V_{oc} is examined for four different pressure values, while keeping $L(0) = 4.0a_b^*$ and $l_2(0) = 1.0a_b^*$. It is evident that V_{oc} exhibits a monotonic decrease with increasing indium concentration. In parallel, a distinct trend emerges wherein V_{oc} increases consistently with applied pressure, regardless of the In-content. Notably, V_{oc} shows an almost linear rise from approximately 1.1 V at P = 0 GPa to 1.6 V at P = 30 GPa. This significant enhancement is primarily attributed to the widening of the band gap induced by the application of hydrostatic pressure, which positively impacts the overall device performance. Nevertheless, it is important to note that despite the pressure-induced improvement, the absolute values of V_{oc} remain relatively low. This limitation is largely due to the detrimental effects of spontaneous and piezoelectric polarization fields, which persist across the range of compositions and pressures studied.

Figure 12 presents the computed results of the photo-generated current density j_{sc} as a function of indium composition. The analysis is performed at room temperature, assuming an on-center impurity configuration with $l_1(0) = 2.0$ nm. Panel (12a) illustrates the behavior of j_{sc} under fixed conditions: P = 10 GPa, $l_2(0) = 1.0$ nm, and four different values of the barrier width L(0). The results indicate that j_{sc} increases almost linearly with indium content up to a certain threshold, beyond which it tends to saturate. Additionally, j_{sc} exhibits only minimal dependence on the variation of L(0), suggesting that barrier width has a limited effect on the short-circuit current density under the given conditions.

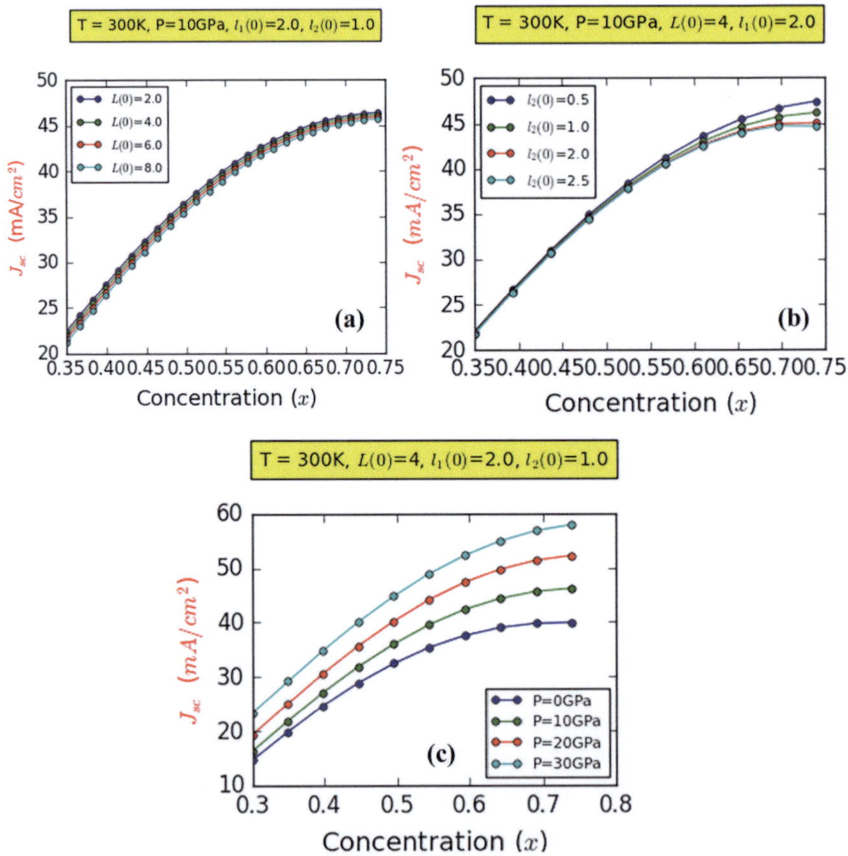

Fig. 12 Variation of the short circuit current (J_{sc}) as a function of the indium-concentration (x) for different values of barrier's width (**a**), InGaN-layer's width (**b**), and hydrostatic pressure (**c**)

This observed trend can be attributed to the intrinsic material property whereby increasing the indium content in $In_xGa_{1-x}N$ results in a concomitant narrowing of the bandgap, which consequently diminishes the optical transition threshold and reduces the energy depth of the InGaN absorber layer. This bandgap reduction enhances the absorption of incident photons by enabling lower-energy transitions, thereby improving photogeneration rates within the intrinsic (i) region. As the indium composition increases, the intrinsic layer becomes more efficient in harvesting incident light and generating charge carriers, thus facilitating higher rates of carrier excitation. It is important to underscore that the presented simulations are limited to cases where no secondary intermediate band (IB) formation is evident. Additionally, the combined influence of spontaneous and piezoelectric polarization effects helps maintain relatively moderate short-circuit current density j_{sc} values throughout the examined parameter space.

Panel (12b) illustrates the variation in short-circuit current density (j_{sc}) under constant parameters: applied pressure P = 10 GPa and effective Bohr radius $L(0) = 1.0a_b^*$. The analysis encompasses four distinct initial widths of the InGaN absorption layer, denoted as $l_2(0)$, in the absence of additional pressure modulation. Across all considered layer widths, an increase in indium composition leads to a monotonic enhancement in j_{sc}. This behavior is primarily attributed to the rise in photogeneration rate, stemming from the indium-induced reduction in bandgap and corresponding decrease in the energy depth of the InGaN layer. These modifications result in lower transition energies, enabling the absorption of a broader photon energy spectrum and, consequently, the generation of a greater number of electron–hole pairs. This leads to the observed increase in j_{sc} with higher indium incorporation. However, it is noteworthy that beyond a critical composition threshold, identified as $x_c = 48\%$, the impact of the right quantum well (RQW) becomes increasingly pronounced, significantly influencing carrier dynamics and optical transitions within the structure.

A reduction in the width of the right quantum well (RQW) amplifies its effectiveness in mitigating the influence of the strong built-in electric field (BEF), thereby highlighting the strategic advantage of incorporating quantum wells within semi-graded structures. To further examine these effects, the influence of hydrostatic pressure on the short-circuit current density (j_{sc}) is analyzed under fixed conditions: an InGaN-layer width of $l_2(0) = 1.0a_b^*$, and a structural length $L(0) = 4.0a_b^*$. As illustrated in Panel (5c), an increase in indium composition consistently yields a rise in j_{sc} across all pressure values considered. Additionally, the data clearly demonstrate that, for a fixed composition, increasing the applied pressure results in a more pronounced enhancement of j_{sc}. This pressure-induced improvement is especially significant at higher indium concentrations. The underlying mechanism can be attributed to the interplay between pressure and photon absorption processes: increasing pressure induces a widening of the material's bandgap, which subsequently raises the energy of absorbed photons. This shift enhances the generation of high-energy carriers, ultimately contributing to the observed increase in j_{sc}.

Figure 13 displays the simulated results for photovoltaic conversion efficiency (η) as a function of indium composition (x), evaluated under the assumption of an on-center impurity configuration at room temperature. The structural parameter is set to $l_1(0) = 2.0a_b^*$. Across all panels and regardless of the specific parameter variations, a prominent and consistent trend emerges: η exhibits a distinct peak at a critical indium composition. This optimal performance is most evident within the composition range $0.5 < x < 0.6$, where in the majority of optical transitions are optimally aligned, resulting in enhanced photon absorption and increased carrier generation. The concurrent improvement in both absorption and current density collectively drives the efficiency to its maximum value. Conversely, for compositions exceeding this critical range, a decline in η is observed, which can be attributed to the degradation of material quality in the high-indium-content InGaN layers—manifested as increased defect density, phase separation, or reduced crystalline integrity—ultimately hindering photovoltaic performance.

The observed degradation in device performance is primarily attributed to the combined effects of spontaneous and piezoelectric polarization fields, as well as the increased density of dislocations arising from lattice mismatch between the constituent layers. These issues become more pronounced as the indium composition surpasses 50%, presenting significant challenges in the growth of high-quality

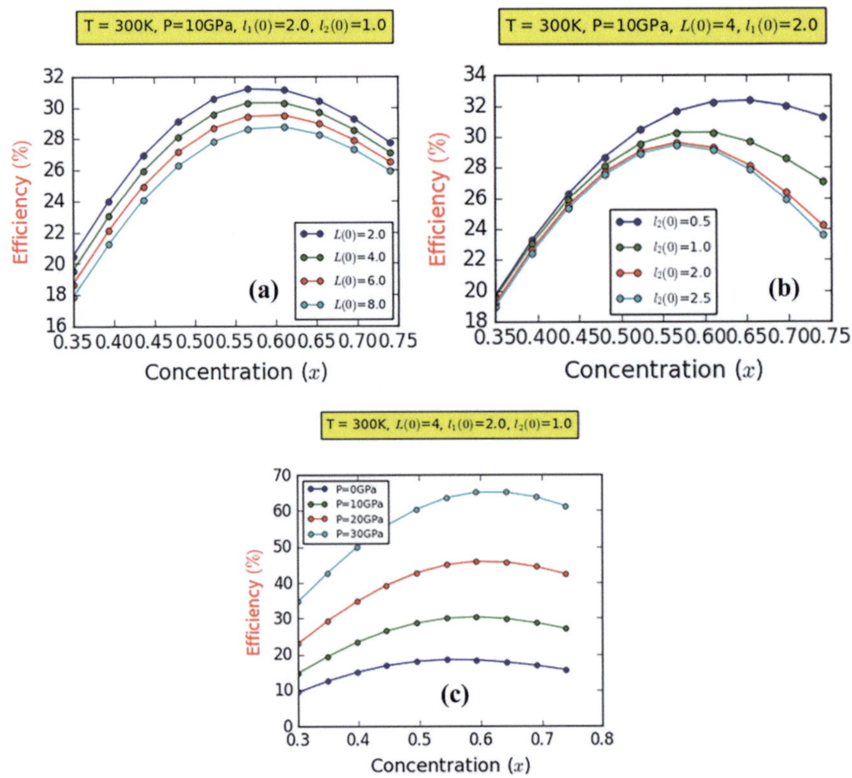

Fig. 13 Variation of the efficiency (%) as a function of the In-concentration for different values of barrier's width (**a**), InGaN-layer's width (**b**), and hydrostatic pressure (**c**)

InGaN layers. Such structural and material constraints ultimately contribute to the decline in photovoltaic efficiency [36].

Panel (13a) investigates the variation in photovoltaic conversion efficiency (η) under an applied hydrostatic pressure of P = 10 GPa, with the geometric condition $l_1(0) = 2.0a_b^*$, across four distinct values of GaN-layer thicknesses. For all examined indium compositions, η exhibits a characteristic behavior—initially increasing to a peak value before decreasing. Furthermore, a notable enhancement in η is observed as the GaN layer thickness L(0) is reduced from $8a_b$* to $2a_b$*. This improvement is attributed to the concurrent rise in both the short-circuit current density j_{sc} and the open-circuit voltage V_{oc}. Specifically, the efficiency increases from 28.63 to 31.22%, representing a relative improvement of approximately 11.5%.

In Panel (13b), the effect of the InGaN-layer thickness on photovoltaic conversion efficiency (η) is investigated under a constant hydrostatic pressure of P = 10 GPa. This analysis is conducted for fixed GaN and InN layer thicknesses of $L(0) = 4a_b^*$ and $l_1(0) = 1.0a_b^*$, respectively. The results highlight the role of absorber layer geometry in modulating η and further emphasize the interplay between structural parameters and polarization effects in determining device performance.

As illustrated in the final panel, a reduction in the thickness of the InGaN layer leads to a noticeable improvement in photovoltaic conversion efficiency (η), with the enhancement being particularly pronounced at higher indium compositions. The effect of InGaN-layer thickness on η closely mirrors that observed with variations in GaN-layer thickness, especially in the low-indium-composition regime.

This behaviour can be attributed to the enhanced photon absorption capabilities of intermediate bands (IBs) when the GaN layer is sufficiently thin. Under such conditions, the IBs facilitate efficient sub-bandgap photon absorption, promoting increased electron excitation and thereby generating a higher short-circuit current density j_{sc}, which ultimately contributes to improved photovoltaic efficiency. However, it is crucial to note that the presence of spontaneous and piezoelectric polarization fields adversely affects device performance by introducing internal electric fields that hinder carrier collection and reduce overall efficiency.

Panel (13c) presents the variation of η as a function of indium composition for four distinct hydrostatic pressure values, under a fixed structural configuration defined by $L(0) = 4a_b^*$ and $l_1(0) = 2.0a_b^*$, at room temperature. The results clearly demonstrate that hydrostatic pressure exerts a substantial influence on the photovoltaic performance of intermediate band solar cells (IBSCs). Specifically, pressure application significantly enhances η, with this effect becoming increasingly pronounced at higher indium concentrations. For example, at a fixed composition of $x = 0.6$, increasing the applied pressure from 0 to 30 GPa leads to a dramatic efficiency increase—from approximately 16 to 65.3%. This underscores the critical role of pressure-induced modifications in band structure and carrier dynamics, particularly in high-indium-content InGaN systems. This translates to an estimated relative efficiency increase of approximately 24.62%. The underlying mechanism can be attributed to the effect of hydrostatic pressure, which promotes enhanced electronic transitions from the valence band to the conduction band. This pressure-induced band structure modification increases carrier generation, thereby elevating the photocurrent and contributing to the observed improvement in the overall efficiency of the intermediate band solar cell (IBSC).

6 Conclusion

This chapter highlights the significant impact of various physical and technological factors on the performance of intermediate band solar cells incorporating quantum wells. The systematic investigation of alloy composition, quantum well confinement, structural dimensions, as well as external influences such as hydrostatic pressure and applied electric and magnetic fields, reveals their critical roles in shaping the electronic and optical properties of these devices. By modulating key photovoltaic parameters—including open-circuit voltage, short-circuit current density, fill factor, and overall efficiency—these factors offer multiple avenues for optimization. A deeper understanding of these interdependencies provides valuable insights that are essential for advancing the design and fabrication of high-efficiency solar cells based on III-nitride materials. Future work should focus on integrating these findings into practical device architectures and exploring additional environmental conditions to further enhance solar cell performance.

References

1. A. Luque, A. Martí, Increasing the efficiency of ideal solar cells by photon induced transitions at intermediate levels. Phys. Rev. Lett. **78**, 5014–5017 (1997). https://doi.org/10.1103/PhysRevLett.78.5014
2. W. Shockley, H.J. Queisser, Detailed balance limit of efficiency of p–n junction solar cells. J. Appl. Phys. **32**, 510–519 (1961). https://doi.org/10.1063/1.1736034
3. Temperature dependence of InGaN/GaN multiple quantum well based high efficiency solar cell. ScienceDirect. Available online: https://www.sciencedirect.com/science/article/abs/pii/S0927024811003898?via%3Dihub. Accessed 30 May 2025
4. R. Belghouthi, M. Aillerie, Temperature dependence of InGaN/GaN multiple quantum well solar cells. Energy Procedia **157**, 793–801 (2019). https://doi.org/10.1016/j.egypro.2018.11.245
5. Temperature dependences of $In_xGa_{1-x}N$ multiple quantum well solar cells. IOPscience. Available online: https://doi.org/10.1088/0022-3727/42/10/105101. Accessed 30 May 2025
6. B. Chouchen, A.E. Aouami, M.H. Gazzah, A. Bajahzar, E.M. Feddi, F. Dujardin, H. Belmabrouk, Modeling the impact of temperature effect and polarization phenomenon on InGaN/GaN-multi-quantum well solar cells. Optik **199**, 163385 (2019). https://doi.org/10.1016/j.ijleo.2019.163385
7. G.F. Brown, J.W. Ager III., W. Walukiewicz, J. Wu, Finite element simulations of compositionally graded InGaN solar cells. Sol. Energy Mater. Sol. Cells **94**, 478–483 (2010)
8. Simulation of high-efficiency GaN/InGaN p-i-n solar cell with suppressed polarization and barrier effects. IEEE J. Mag. IEEE Xplore. Available online: https://ieeexplore.ieee.org/document/6334407. Accessed 30 May 2025
9. K.-H. Lee, P.-C. Chang, S.-J. Chang, AlGaN/GaN high electron mobility transistors based on InGaN/GaN multi-quantum-well structures with photo-chemical vapor deposition of SiO_2

dielectrics. Microelectron. Eng. **104**, 105–109 (2013). https://doi.org/10.1016/j.mee.2012.11.020
10. J.F. Muth, J.H. Lee, I.K. Shmagin, R.M. Kolbas, H.C. Casey Jr., B.P. Keller, U.K. Mishra, S.P. DenBaars, Absorption coefficient, energy gap, exciton binding energy, and recombination lifetime of GaN obtained from transmission measurements. Appl. Phys. Lett. **71**, 2572–2574 (1997). https://doi.org/10.1063/1.120191
11. Z. Yang, Z. Xu, A theoretical study of electronic and optical properties in wurtzite GaN. J. Phys. Condens. Matter **8**, 8303 (1996). https://doi.org/10.1088/0953-8984/8/43/022
12. R. En-nadir, H. El-ghazi, A. Jorio, I. Zorkani, H. Abboudi, F. Jabouti, Numerical study of temperature and electric field effects on the total optical absorption coefficient in the presence of optical inter-conduction-subband transitions in InGaN/GaN single parabolic quantum wells. FDMP **18**, 1253–1261 (2022). https://doi.org/10.32604/fdmp.2022.021759
13. H.M. Baghramyan, M.G. Barseghyan, A.A. Kirakosyan, R.L. Restrepo, M.E. Mora-Ramos, C.A. Duque, Donor impurity-related linear and nonlinear optical absorption coefficients in $GaAs/Ga_{1-x}Al_xAs$ concentric double quantum rings: effects of geometry, hydrostatic pressure, and aluminum concentration. J. Lumin. **145**, 676–683 (2014). https://doi.org/10.1016/j.jlumin.2013.08.061
14. J.F. Muth, J.D. Brown, M.A.L. Johnson, Z. Yu, R.M. Kolbas, J.W. Cook Jr., J.F. Schetzina, Absorption coefficient and refractive index of GaN, AIN and AlGaN alloys. MRS Online Proc. Libr. (OPL) **537**, G5.2 (1998). https://doi.org/10.1557/PROC-537-G5.2
15. Y.P. Varshni, Temperature dependence of the energy gap in semiconductors. Physica **34**, 149–154 (1967). https://doi.org/10.1016/0031-8914(67)90062-6
16. H. Abboudi, R. En-nadir, M.B. Kabatas, Tuning intermediate band solar cell efficiency: the interplay of electric fields, composition, impurities, and confinement. Available online: https://doi.org/10.3390/nano14221858
17. V. Barsan, A new analytic approximation for the energy eigenvalues of a finite square well. Rom. Rep. Phys. **64**, 685–694 (2012)
18. W. Shockley, The Shockley–Queisser limit. J. Appl. Phys. **32**, 510–519 (1961)
19. W. van Roosbroeck, W. Shockley, Photon-radiative recombination of electrons and holes in germanium. Phys. Rev. **94**, 1558–1560 (1954). https://doi.org/10.1103/PhysRev.94.1558
20. D. Qing-Wen, W. Xiao-Liang, Y. Cui-Bai, Computational investigation of $In_xGa_{1-x}N/InN$ quantum-dot intermediate-band solar cell. Chin. Phys. Lett. **28**, 018401 (2011). https://doi.org/10.1088/0256-307X/28/1/018401
21. M.A. Green, Solar cell fill factors: general graph and empirical expressions. Solid-State Electron. **24**, 788–789 (1981). https://doi.org/10.1016/0038-1101(81)90062-9
22. D. Qingwen, W. Xiaoliang, X. Hongling, M. Zeyu, Z. Xiaobin, H. Qifeng, L. Jinmin, W. Zhanguo, Theoretical investigation of efficiency of a Pa-SiC: H/Ia-Si: H/n-Mc-Si solar cell. J. Semicond. **31**, 103003 (2010)
23. S.-H. Park, Crystal orientation effects on electronic properties of wurtzite GaN/AlGaN quantum wells with spontaneous and piezoelectric polarization. Jpn. J. Appl. Phys. **39**, 3478 (2000). https://doi.org/10.1143/JJAP.39.3478
24. A. Luque, A. Martí, The intermediate band solar cell: progress toward the realization of an attractive concept. Adv. Mater. **22**, 160–174 (2010). https://doi.org/10.1002/adma.200902388
25. H. El Ghazi, A. Jorio, Electron-hole transition in spherical QD-QW nanoparticles based on GaN|(In,Ga)N|GaN under hydrostatic pressure. Phys. B **429**, 42–45 (2013). https://doi.org/10.1016/j.physb.2013.07.025
26. H. El Ghazi, A.J. Peter, Threshold pump intensity effect on the refractive index changes in InGaN SQD: internal constitution and size effects. Phys. B **462**, 30–33 (2015). https://doi.org/10.1016/j.physb.2015.01.014
27. R. En-nadir, M.A. Basyooni-M. Kabatas, M. Tihtih, W. Belaid, I. Ez-Zejjari, E.G. Majda, H. El Ghazi, A. Sali, I. Zorkani, Enhancing emission via radiative lifetime manipulation in ultrathin InGaN/GaN quantum wells: the effects of simultaneous electric and magnetic fields, thickness, and impurity. Nanomaterials **13**, 2817 (2023). https://doi.org/10.3390/nano13212817

28. G.F. Brown, J.W. Ager, W. Walukiewicz, J. Wu, Finite element simulations of compositionally graded InGaN solar cells. Sol. Energy Mater. Sol. Cells **94**, 478–483 (2010). https://doi.org/10.1016/j.solmat.2009.11.010
29. O. Ambacher, Growth and applications of group III-nitrides. J. Phys. D Appl. Phys. **31**, 2653 (1998). https://doi.org/10.1088/0022-3727/31/20/001
30. T. Takeuchi, S. Sota, M. Katsuragawa, M. Komori, H. Takeuchi, H.A.H. Amano, I.A.I. Akasaki, Quantum-confined stark effect due to piezoelectric fields in GaInN strained quantum wells. Jpn. J. Appl. Phys. **36**, L382 (1997). https://doi.org/10.1143/JJAP.36.L382
31. J. Shi, Z. Gan, Effects of piezoelectricity and spontaneous polarization on localized excitons in self-formed InGaN quantum dots. J. Appl. Phys. **94**, 407–415 (2003). https://doi.org/10.1063/1.1576490
32. B.K. Ridley, W.J. Schaff, L.F. Eastman, Theoretical model for polarization superlattices: energy levels and intersubband transitions. J. Appl. Phys. **94**, 3972–3978 (2003). https://doi.org/10.1063/1.1601686
33. F. Bernardini, V. Fiorentini, Spontaneous versus piezoelectric polarization in III–V nitrides: conceptual aspects and practical consequences. Phys. Status Sol. (b) **216**, 391–398 (1999). https://doi.org/10.1002/(SICI)1521-3951(199911)216:1<391::AID-PSSB391>3.0.CO;2-K
34. O. Ambacher, J. Majewski, C. Miskys, A. Link, M. Hermann, M. Eickhoff, M. Stutzmann, F. Bernardini, V. Fiorentini, V. Tilak et al., Pyroelectric properties of Al(In)GaN/GaN hetero- and quantum well structures. J. Phys. Condens. Matter **14**, 3399 (2002). https://doi.org/10.1088/0953-8984/14/13/302
35. V. Fiorentini, F. Bernardini, O. Ambacher, Evidence for nonlinear macroscopic polarization in III–V nitride alloy heterostructures. Appl. Phys. Lett. **80**, 1204–1206 (2002)
36. M.F. Zscherp, S.A. Jentsch, M.J. Müller, V. Lider, C. Becker, L. Chen, M. Littmann, F. Meier, A. Beyer, D.M. Hofmann et al., Overcoming the miscibility gap of GaN/InN in MBE growth of cubic $In_xGa_{1-x}N$. ACS Appl. Mater. Interfaces **15**, 39513–39522 (2023). https://doi.org/10.1021/acsami.3c06319

Recombination Dynamics and Radiative Lifetimes in III–N Semiconductor Photonic Structures

Radouane En-Nadir

1 Introduction

The recombination process and associated radiative lifetime (RT) play a pivotal role in determining the efficiency and functionality of nano- and micro-scale optoelectronic devices that utilize low-dimensional quantum well (QW) structures [1–3]. These parameters govern the duration for which excited electrons remain in an elevated energy state before recombining and emitting photons, thereby directly impacting the emission efficiency and response speed of optoelectronic devices. In the context of InGaN-based quantum wells, shorter radiative lifetimes are particularly advantageous, as they facilitate efficient photon emission—a key requirement for high-performance LEDs and laser diodes [4]. Consequently, precise control over the radiative lifetime is essential for optimizing the performance of quantum well-based devices across a wide range of applications, including solar cells, light-emitting diodes (LEDs), laser sources, photodetectors, and optical modulators [5–8]. Tuning the radiative lifetime of electron–hole pairs on ultrafast timescales—ranging from picoseconds to femtoseconds—has a direct impact on both the emission rate and overall photon generation efficiency [9]. In devices like LEDs and laser diodes, the radiative lifetime is a critical determinant of photon emission efficiency. Reduced radiative lifetimes enable quicker recombination of electron–hole pairs, resulting in more rapid and efficient light generation [10]. Conversely, extended radiative lifetimes can increase the likelihood of non-radiative recombination pathways, where energy is dissipated without photon emission [11], In such cases, the excess energy is lost as heat, leading to a reduction in the device's luminous efficiency. Quantum wells mitigate this by confining charge carriers within two dimensions, which enhances the overlap of electron and hole [9]. For example, InGaN/GaN quantum wells

R. En-Nadir (✉)
Institut Interdisciplinaire d'Innovation Technologique (3IT), Université de Sherbrooke, 3000 boulevard Université, Sherbrooke, Québec J1K 0A5, Canada
e-mail: radouane.en-nadir@usherbrooke.ca

R. En-nadir (ed.), *Quantum Systems at the Nanoscale*, Lecture Notes in Nanoscale Science and Technology 45, https://doi.org/10.1007/978-3-032-06645-9_9

(QWs) serve as fundamental building blocks in contemporary optoelectronic devices, including LEDs and laser diodes. By adjusting the indium content, their emission can be precisely tuned across a broad spectral range—from near-infrared to near-ultraviolet—enabling high-efficiency lighting and full-spectrum display technologies [12, 13]. These structures provide robust carrier confinement and high quantum efficiency, resulting in bright and energy-efficient operation. Beyond lighting applications, InGaN/GaN quantum wells are increasingly utilized in solar cells, photodetectors, and cutting-edge micro-display technologies [14, 15]. The optical properties of these systems were intensively investigated due to their impressive characteristics [14–21]. Recent advancements have facilitated their integration into high-resolution micro-LED displays, which are essential components for augmented and virtual reality devices [22] Additionally, their strong photoabsorption and efficient carrier extraction make them ideal for high-performance photodetectors [14]. In photovoltaic applications, multi-quantum well structures are being investigated to improve the efficiency of tandem and concentrator solar cells [23]. In solar cells, short radiative lifetimes promote radiative recombination over non-radiative processes, which can lower the external quantum efficiency and consequently reduce both the output voltage and current [24]. In high-speed optoelectronic devices such as modulators and photodetectors, shorter radiative lifetimes enable quicker carrier recombination, resulting in faster response times and improved switching speeds and detection efficiencies [25]. However, in quantum well laser diodes, shorter radiative lifetimes enhance photon generation efficiency, lowering the threshold current, reducing energy consumption, and minimizing heat production [26]. Furthermore, in quantum wells, carrier confinement increases carrier densities and enhances wave function overlap, which shortens radiative lifetimes and enables tailored designs to optimize device speed or light emission [27]. Nanostructures form the foundation of modern optoelectronic technologies, with ongoing advances in material growth, device engineering, and theoretical modeling driving improvements across lighting, displays, and sensing applications. Central to further progress is a comprehensive understanding of radiative lifetime, a critical yet often underrepresented parameter that governs both efficiency and response speed. This chapter provides an in-depth investigation of radiative lifetimes in heterostructures—particularly II–N materials—filling a notable gap in existing literature. The influence of factors such as layer thickness, temperature, indium composition, defects, and applied electric and magnetic fields on radiative recombination dynamics is systematically examined. By clarifying these dependencies, this work supports the optimized design of ultrafast, multifunctional optoelectronic devices, including high-speed communication systems, energy-efficient lighting, advanced photovoltaic converters, and next-generation sensing platforms spanning the near-infrared to ultraviolet spectral range.

2 Theoretical Framework and Models

For visual clarity and to reference both the ISB and BTB optical transitions, a comprehensive schematic of the studied system is presented in Fig. 1. Investigating the radiative lifetimes of these transitions requires numerical solutions of the one-dimensional Schrödinger equation, which describes the behavior of electrons and holes within the structure. This numerical approach is necessary due to the presence of an impurity term and a finite confinement profile, rendering an analytical solution unfeasible. By solving the equation within the framework of effective-mass theory, the energy levels and corresponding wave functions for electrons in the conduction band and holes in the valence band are obtained. Subsequently, the transition energies and dipole matrix elements (DME) between electron and heavy-hole states are determined. A detailed explanation of the theoretical models is provided in Ref. [28].

The analytical expression of the radiative lifetime is given as follows [29–31]:

$$\tau_{fi} = \frac{3c^3 h^4 \varepsilon_0}{16\pi^3 \Delta E_{fi}^3 n_r \mathcal{M}_{if}^2} \tag{1}$$

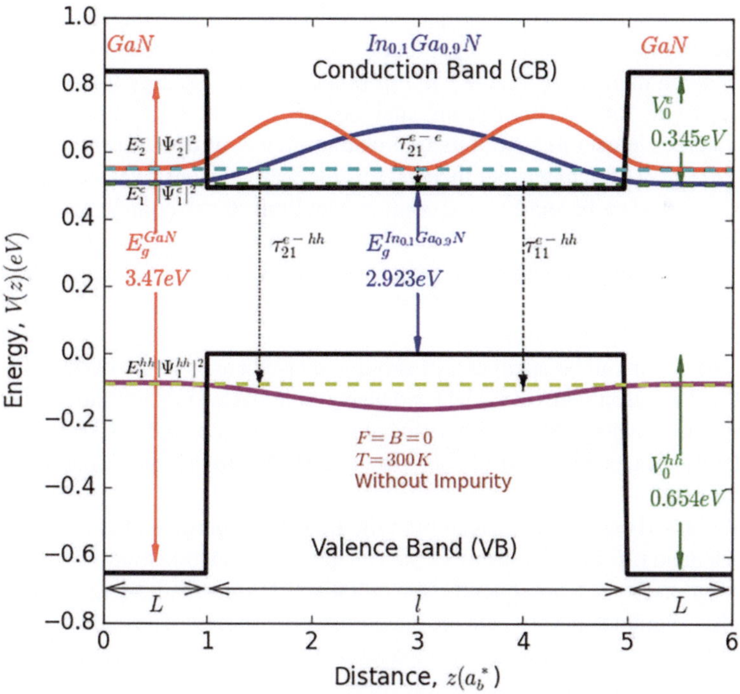

Fig. 1 Schematic representation of a single GaN/InGaN/GaN quantum well illustrating various potential optical transitions, their corresponding energy levels, and associated probability density functions

where c, h, ε_0, n_r denote respectively the speed of light in vacuum, Planck's constant, vacuum permittivity and refractive index relative to the materials used in the study.

For clarity and brevity throughout the following results and figure discussions, abbreviations will be used: LSP for left sub-panel, RSP for right sub-panel, and MP for main panel. For physical parameters and models, refer to Ref. [28].

2.1 Analyses of the Thickness Effect

Figure 2 presents how the transition energy (ΔE_{fi}), the dipole matrix element (M_{if}^2), and the recombination lifetime (τ_{fi}^i, where i represents transitions such as e → e and e → hh) vary with well thickness for three different optical transitions, keeping the barrier thickness fixed at L = 2 and without any external excitations. It is clear that the width of the quantum well strongly influences ΔE_{fi} (shown in the left sub-panel), M_{if}^2 (right sub-panel), and τ_{fi}^i (main panel). For all transitions, increasing the well thickness results in a rise in ΔE_{fi} until it eventually saturates in the weak confinement regime (beyond l = 0.5 nm). This effect is notably stronger for band-to-band (BTB) transitions compared to intersubband (ISB) ones. The underlying reason is that a wider well reduces quantum confinement, which increases the energy difference between excited (E_f) and ground (E_i) states, thereby boosting ΔE_{fi}. Within the strong confinement regime (l < 0.5 nm), the transition energy for BTB remains nearly constant for both transitions analyzed (E_{11}^{e-hh} and E_{21}^{e-hh}). In contrast, the dipole matrix element M_{if}^2 exhibits two distinct trends: it decreases when the well is thin (strong confinement) but increases as the well thickens (weak confinement). This behavior arises because strong confinement limits the overlap between electron wave functions for initial and final states, whereas in weak confinement, this overlap improves. Additionally, within the weak confinement regime (l > 0.5 nm), the dipole matrix element associated with BTB transitions (specifically M_{11}^{e-hh} and M_{21}^{e-hh}) remains essentially unchanged.

This behavior can be attributed to simultaneous changes in the overlap between the relevant electron states ($1s^e$, $2s^e$) and heavy-hole states ($1s^{hh}$) within this confinement regime. Notably, the overlap associated with intersubband (ISB) transitions is consistently greater than that of band-to-band (BTB) transitions, as shown in the right sub-panel.

As discussed earlier, changes in both the transition energy (ΔE_{fi}) and the dipole matrix element (M_{if}) have a strong impact on the recombination lifetime for both ISB and BTB transitions. Consequently, the radiative lifetime exhibits significant variation as these parameters vary with the quantum well thickness. The main panel clearly shows that the radiative lifetime decreases as the well width increases, eventually leveling off for thicker wells. This decline happens more rapidly for ISB transitions compared to BTB transitions. This trend arises from the simultaneous increase

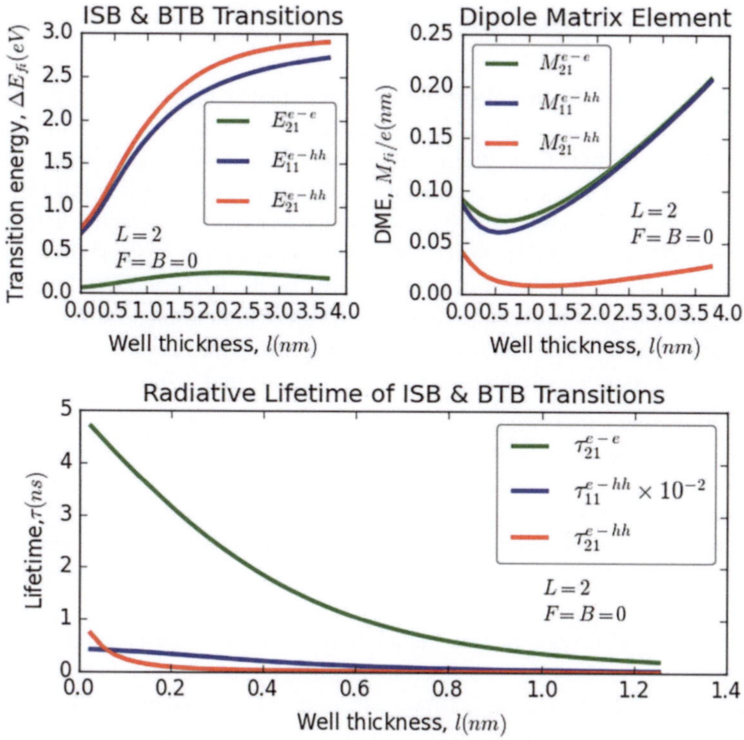

Fig. 2 Variation of transition energy, dipole matrix elements and the recombination lifetime versus the well width for different optical transitions: intra-subband (ISB) and band-to-band (BTB) transitions in InGaN/GaN ultrathin-layered heterostructure (QW)

in both ΔE_{fi} and M_{if} with increasing well thickness, consistent with the explanations given for the left and right sub-panels. From a quantum mechanical perspective, notable differences emerge in radiative behavior depending on the confinement regime, particularly for electron–electron (e-e) transitions. Under strong confinement, the radiative lifetime τ_{fi}^{e-e} is relatively long, around 4.7 ns. In contrast, in the weak confinement regime, this lifetime shortens dramatically to approximately 0.3 ns. For electron-heavy hole (e-hh) transitions, the radiative lifetimes τ_{fi}^{e-hh} are considerably faster, ranging from about 0.8 ns down to nearly instantaneous emission, effectively approaching zero. These pronounced differences highlight how confinement regimes profoundly shape the radiative properties of quantum systems. These findings are in strong agreement with recent studies reporting on the radiative lifetimes of neutral excitons confined in CdSe colloidal nanoplatelets (NPLs) [32], thin GaAs-based quantum box (QB) [33], and GaAs/GaAlAs quantum well nanostructures at room temperature, measured using the photoluminescence phase shift method [34].

2.2 Applied Electric Field Effect

Figure 3 shows how the transition energy, dipole matrix elements, and recombination lifetime depend on the strength of an externally applied electric field, focusing on various optical transitions within an ultrathin InGaN/GaN heterostructure. The transitions analyzed include both intersubband (ISB) and band-to-band (BTB) types. From the left and right sub-panels, it is evident that changes in the electric field magnitude affect ISB and BTB transitions differently, independent of layer thickness and in the absence of any magnetic field. Specifically, for the E_{11}^{e-hh} and E_{21}^{e-hh} transitions, the transition energies decrease linearly as the electric field increases. Conversely, the transition energy E_{21}^{e-e} shows a gradual increase with increasing electric field strength. The dipole matrix elements (M_{ij}) display varied responses depending on the transition: M_{21}^{e-e} and M_{11}^{e-hh} decrease in magnitude with stronger electric fields, whereas M_{21}^{e-hh} increases. This behavior indicates that the overlap between electron and hole wave functions—an essential factor for radiative recombination—can either improve or diminish depending on whether the transition is ISB or BTB. These shifts in transition energies and modifications in electron–hole correlations directly impact the radiative lifetime of the transitions studied, as clearly shown in the main panel of Fig. 3.

The main panel reveals that as the applied electric field strength increases, the radiative lifetime associated with electron–electron correlation (τ_{21}^{e-e}) initially shows a slight rise before gradually declining. In contrast, the radiative lifetime linked to electron-heavy hole correlation (τ_{11}^{e-hh}) steadily increases with the electric field, while τ_{21}^{e-hh} exhibits a notably accelerated increase. Specifically, τ_{21}^{e-e} reaches its maximum near an electric field of approximately $F = 0.37$, marking a critical point where its behavior shifts. This trend results from the competing effects of the transition energy (E_{21}^{e-e}), which increases with the field, and the dipole matrix element (M_{21}^{e-e}), which decreases. Below F = 0.37, variations in E_{21}^{e-e} primarily govern τ_{21}^{e-e}, whereas above this threshold, changes in M_{21}^{e-e} dominate its behavior. The steady increase of both τ_{21}^{e-hh} and τ_{11}^{e-hh} across all field strengths can be largely attributed to the dominant influence of transition energies (E_{21}^{e-hh} and E_{11}^{e-hh}) over the overlap effects (M_{21}^{e-hh} and M_{11}^{e-hh}). Notably, τ_{21}^{e-hh} accelerates more rapidly than τ_{11}^{e-hh}, reflecting the pronounced growth of M_{21}^{e-hh} with increasing electric field intensity. This enhancement in the dipole matrix element has a greater impact on τ_{21}^{e-hh}, explaining the differences observed in their responses. Additionally, τ_{11}^{e-hh} consistently exhibits shorter radiative lifetimes compared to the other transitions studied. Radiative lifetimes for (e-e) transitions fall roughly between 0.20 and 0.25 ns. In contrast, (e-hh) transition lifetimes range from approximately 0.13 ns down to nearly instantaneous emissions, approaching zero nanoseconds under weak electric fields. This clear distinction underscores the significant role of applied electric fields in modulating recombination dynamics within the quantum system. These results are in strong agreement with recent literature examining the influence of electric fields on recombination lifetimes in nanostructures, such as GaAs-doped quantum dots [29].

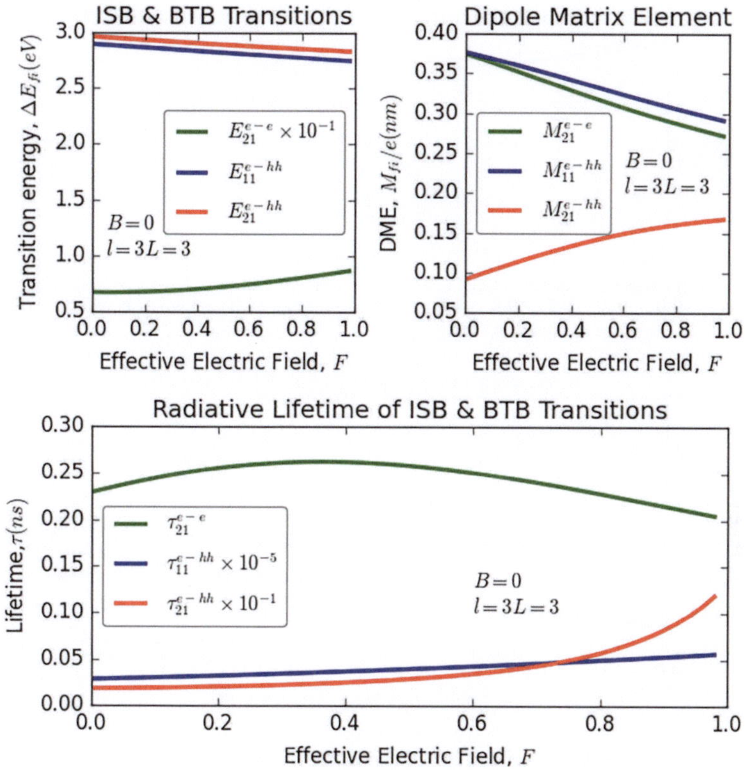

Fig. 3 Variation of transition energy, dipole matrix elements and the recombination lifetime versus the applied electric field for different optical transitions: intra-subband (ISB) and band-to-band (BTB) transitions in InGaN/GaN ultrathin-layered heterostructure (QW)

2.3 *Applied Magnetic Field Effect*

Figure 4 highlights the complex influence of magnetic field strength on key optical properties—namely transition energy, dipole matrix elements, and recombination lifetime—focusing on both ISB and BTB transitions. The figure clearly shows that increasing the magnetic field significantly affects the transition energy difference (ΔE_{fi}), dipole matrix elements (M_{if}), and recombination lifetime (τ_{fi}) across the optical transitions studied. As the magnetic field strengthens, the transition energies E_{21}^{e-hh} and E_{11}^{e-hh} decrease. This behavior results from the widening energy separation between levels E_1^e, E_2^e, and E_1^{hh} caused by the magnetic field's influence. Correspondingly, the dipole matrix elements M_{21}^{e-e} and M_{11}^{e-hh} also decrease, reflecting a reduced overlap between the associated wave functions under the magnetic field. In contrast, the transition energy E_{21}^{e-e} increases with the magnetic field, demonstrating a distinct response compared to the other transitions. These observations illustrate the nuanced interplay between magnetic field strength and the optical response of the system.

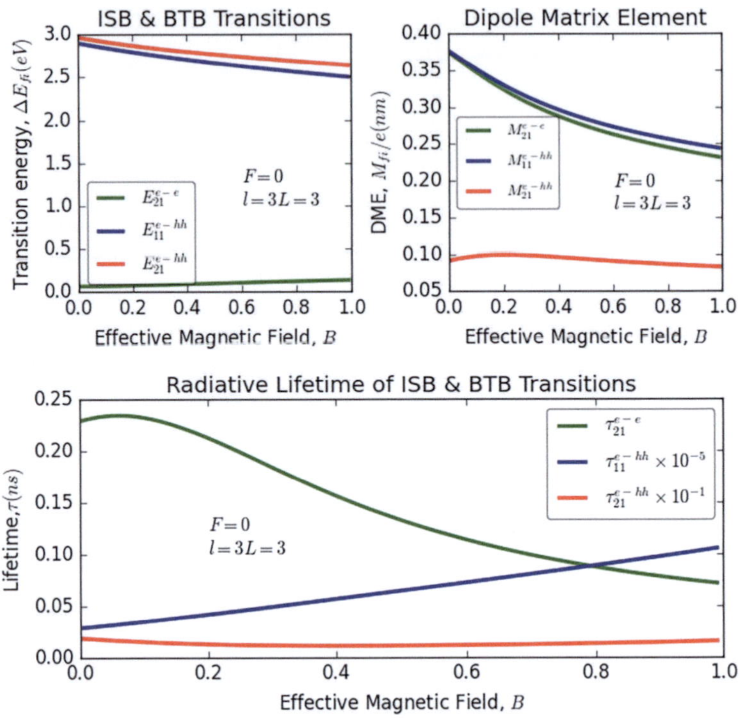

Fig. 4 Variation of transition energy, dipole matrix elements and the recombination lifetime versus the magnetic field for different optical transitions: intra-subband (ISB) and band-to-band (BTB) transitions in InGaN/GaN ultrathin-layered heterostructure (QW)

Interestingly, the dipole matrix element related to the electron-heavy hole transition shows two different trends: it increases when the magnetic field is low (below 0.2) but then gradually decreases as the magnetic field becomes stronger. These changes, which depend on the magnetic field strength, help explain the observed patterns in the data. These variations have a clear impact on the radiative lifetimes in the system, as shown in the main panel of Fig. 4. As the magnetic field gets stronger, the recombination lifetime for electron–electron transitions decreases noticeably. Meanwhile, the recombination lifetime for one of the electron-heavy hole transitions increases significantly, while the other electron-heavy hole transition experiences only a small change. Among these, the shortest lifetime belongs to the particular electron-heavy hole transition that shows the significant increase. The electron–electron recombination lifetime varies roughly between 0.05 and 0.24 ns, reaching its highest value at weak magnetic fields. In contrast, the recombination lifetimes related to electron-heavy hole transitions stay very close to zero throughout the range of magnetic fields tested. This clear difference in how the lifetimes behave highlights the strong effect that magnetic fields have on recombination dynamics within this quantum system.

These findings agree well with recent studies examining how applied fields influence recombination lifetimes in nanostructures [35].

2.4 Defect and Impurity-Location Effect

The results compiled in Table 1 provide a clear depiction. This table presents how variations in impurity position within a thin quantum well—without any external electromagnetic excitation and under a constant nanostructure configuration—affect key optical properties. It is evident that the impact of impurity on the transition energy difference ΔE_{fi} and the dipole matrix element M_{fi} is minor, though not negligible, with changes on the order of 10^{-4}. Consequently, the modulation of the recombination lifetime τ_{fi} remains limited. Initially, the impurity was placed at the left interface of the quantum well and gradually shifted towards its center. This relocation slightly alters the energy levels of the participating states. Moreover, the overlap between the wave functions is also influenced. The matrix element initially decrease as the impurity moves away from the interface but begin to increase as it approaches the center of the well. This trend reflects the role of electron-impurity interaction, which is stronger near the barrier/well interfaces, where it tends to suppress the spatial overlap between electron and hole wave functions. Such reduced overlap leads to a decrease in radiative recombination efficiency in those regions.

Table 1 The effect of impurity-location on the transition energy, dipole matrix element and lifetime

$l = 3L = 3a_b^*$ F = B = 0 z_0 (a_b^*)	E_{21}^{e-e} (eV)	E_{11}^{e-hh} (eV)	E_{21}^{e-hh} (eV)
0.0	0.06625	2.89760	2.96386
0.5	0.06610	2.89764	2.96375
1.0	0.06604	2.89767	2.96371
1.5	0.06608	2.89766	2.96374
z_0 (a_b^*)	M_{21}^{e-e} (nm)	M_{11}^{e-hh} (nm)	M_{21}^{e-hh} (nm)
0.0	0.37256	0.37511	0.09271
0.5	0.37224	0.37489	0.09291
1.0	0.37221	0.37486	0.09287
1.5	0.37253	0.37507	0.09254
z_0 (a_b^*)	τ_{21}^{e-e} (ns)	$\tau_{11}^{e-hh} \times 10^{-6}$ (ns)	τ_{21}^{e-hh} (ns)
0.0	0.24382	2.8755	0.00170
0.5	0.24591	2.8787	0.00169
1.0	0.24667	2.8791	0.00167
1.5	0.24578	2.8760	0.00164

Nevertheless, it is important to highlight that the impurity relocation has no observable effect on the dipole matrix element related to ISB and BTB, which continues to increase steadily as the impurity is moved toward the center of the well. In contrast, the positioning of the impurity does influence the transition energies, dipole matrix elements, and the electron–heavy hole correlations, thereby affecting the radiative lifetimes associated with the intersubband (ISB) and band-to-band (BTB) transitions analyzed in this work, as shown in the third section of Table 1. The data reveal that radiative lifetime between ISB and BTB initially increase slightly before decreasing. This behavior stems from a non-monotonic variation in both the transition energies and their corresponding dipole matrix elements. Specifically, the initial enhancement is due to the increasing overlap and energy difference, while the later decline reflects the diminishing overlap as the impurity approaches the central region. In contrast, the radiative lifetime shows a gradual increase with impurity displacement toward the center. This trend is explained by a simultaneous rise in both energy difference and dipole matrix elements, indicating a favorable condition for enhanced radiative recombination in this configuration.

Figure 5 presents the evolution of electron (E_i^e) and hole (E_i^h) energy levels as a function of quantum well width, considering variations in indium (In) content and temperature. The results are organized into six panels: panels (a), (b), and (c) correspond to different In-fractions, while panels (d), (e), and (f) illustrate the effect of temperature on the same energy levels. Across all panels, a consistent trend is observed—both (E_i^e) and hole (E_i^h) decrease as the well width increases. This behavior holds true regardless of the indium composition or the operating temperature. Such a trend can be attributed to the reduction in quantum confinement with increasing well width, allowing the energy levels to shift closer to the band edges. These findings align well with those reported in the literature [36]. This behavior is predominantly governed by the quantum confinement effects intrinsic to quantum well (QW) structures. In narrower wells, the spatial confinement of charge carriers is stronger, resulting in elevated energy levels due to the limited freedom of motion. As the well width increases, the confinement potential diminishes, thereby lowering the energy levels of both electrons and holes. Notably, panels (b) and (e) reveal a distinctive trend concerning the electron's first excited state energy. In relatively thick quantum wells, where strong confinement effects are still significant (specifically for $l < 0.1$ nm), this excited state energy remains nearly constant. This can be attributed to the enhanced localization of the electron in its excited state compared to the ground state, which stabilizes the energy level despite the increasing well width.

The influence of indium composition on the energy levels is clearly evident in the results. As the indium content in the well material (InGaN) increases, the energy levels of both electrons and holes exhibit an upward shift. This elevation arises from enhanced quantum confinement, which is strengthened by the increased indium concentration. The higher indium fraction contributes to a larger effective mass for the charge carriers, thereby improving their spatial localization within the quantum well. As a result, carriers are subjected to a deeper potential well, leading to elevated quantized energy levels. On the other hand, temperature exerts an opposing effect: as the temperature rises, the energy levels of both electrons and holes tend to decrease. This

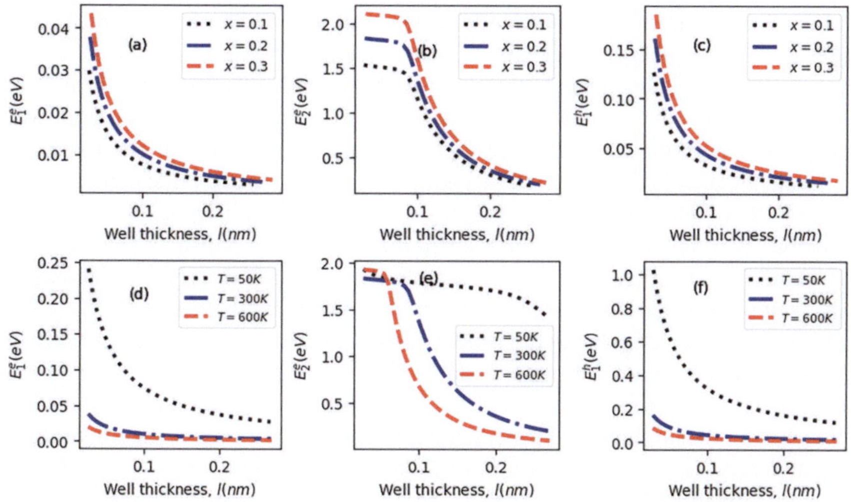

Fig. 5 Representation of electron (E_i^e) and hole (E_i^h) energy variation as a function of well width for different indium (In) fractions and temperatures. Panels **a**–**c** show the energy levels as the well width varies for three values of In-fraction, while panels **d**–**f** illustrate the energy variations for three distinct temperatures. In each case, i represents the corresponding energy level index

behavior aligns well with established findings in the literature, confirming the thermally induced bandgap narrowing and reduced confinement at elevated temperatures [34, 35]. This behavior can be attributed to the increased carrier mobility with rising temperature, which elevates the kinetic energy of charge carriers. However, higher temperatures also induce more frequent scattering events, potentially altering the carriers' energy states. Quantitatively, the ground state electron energy E_1^e varies from approximately 0.043 to nearly 0.005 eV with increasing indium composition, while temperature variations result in a broader shift from about 0.25 to 0.01 eV. For the first excited electron state E_2^e, the energy spans from roughly 2.2 to 0.1 eV as a function of indium content, and from 1.9 to approximately 0.08 eV with temperature changes. Similarly, the ground state hole energy E_1^h ranges from 0.18 to 0.01 eV with varying indium fraction, and from 1 to 0.1 eV with temperature. These results indicate that indium composition has a stronger effect on the electron energy levels, while temperature exerts a more pronounced influence on hole energy. This distinction highlights the pivotal role of material composition and thermal environment in tailoring the electronic properties of quantum well (QW) structures. Consequently, precise optimization of the indium fraction and thermal management is essential to enhance the performance of optoelectronic devices relying on InGaN/GaN QWs. These factors critically impact key parameters such as transition energy, dipole matrix elements (which determine transition probabilities), and radiative lifetimes—parameters that ultimately govern device efficiency.

Figure 6 illustrates the dependence of quantum well width on key optical properties—transition energy (TE), dipole matrix element (DME), and radiative lifetime (RLT)—for intersubband (ISB) and band-to-band (BTB) transitions at three different indium (In) compositions. Panels (a)–(c) show how TE varies with well width, emphasizing the role of In-fraction. In panels (d)–(f), the variation of DME is presented, quantifying the electron-photon coupling strength, while panels (g)–(i) depict RLT, which characterizes how long electrons remain in excited states before recombination. In the first row, the transition energies exhibit distinct behaviors for ISB and BTB transitions. For ISB transitions, TE decreases sharply with increasing well width, particularly in the regime of strong quantum confinement (well width < 0.5 nm). This drop is due to the weakening of confinement, which causes energy levels to shift downward. The ground-state energy of the hole decreases more rapidly than that of the electron's ground state but less rapidly than that of the electron's first excited state, illustrating the differential confinement sensitivity of each state. BTB transition energies, by contrast, show noticeable changes only in narrower wells and remain relatively stable in wider wells. This stability is attributed to the dominant contribution of the band gap energy—typically spanning a few eV—to BTB transitions, which diminishes the relative influence of quantum confinement compared to the meV-scale subband levels. Increasing the In-fraction also influences TE: ISB transition energy increases with In-content, while BTB transition energy decreases. This effect is more pronounced in ISB transitions (as shown in panels b and c) due to their strong confinement dependence. In contrast, BTB transitions are more influenced by the intrinsic band gap of the material, which decreases with higher In content in accordance with Vegard's law. Specifically, ISB transition energies vary from ~ 0.01 to ~ 0.21 eV, while BTB transitions span ~ 2.1 to ~ 3.1 eV, reflecting combined contributions from subband states and band structure. These findings underscore the importance of individual quantum states and material composition in tailoring the optical transitions of quantum well systems.

Moreover, as shown in panels (d), (e), and (f) of Fig. 6, the dipole matrix elements (DME) associated with both intersubband (ISB) and band-to-band (BTB) transitions display a strong dependence on quantum well thickness and indium composition (In-fraction). Notably, the DME increases quasi-linearly with well thickness. This behavior arises from the broader spatial extension of the electron and hole wave functions in wider wells, which enhances the wave function overlap and thus increases the transition probability between quantum states. In contrast, increasing the In-fraction results in a marked decrease in the DME. This reduction is especially pronounced for BTB transitions involving the first excited state of the electron (E_2^e) and the ground state of the hole (E_1^h). This behavior can be explained by the effect of indium incorporation on the effective bandgap of the $In_xGa_{1-x}N$ alloy. A higher In content reduces the conduction and valence band separation, thereby lowering the transition energy and affecting the dipole matrix element, particularly in BTB processes. For ISB transitions, the influence of In-fraction on the DME is comparatively weaker. This is because the strong quantum confinement in narrow wells restricts the spatial redistribution of electron states, making the DME relatively insensitive to composition

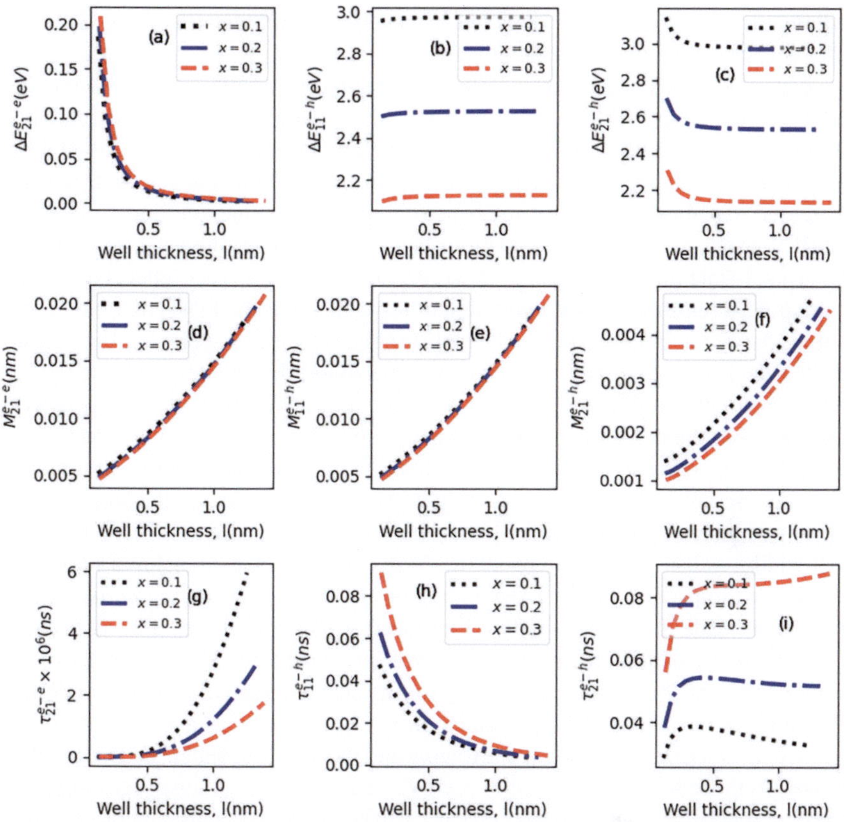

Fig. 6 Illustration of intra-subband and band-to-band optical energy transitions (ISB/BTB), dipole matrix elements (DME), and radiative lifetimes (RLT) of electrons as functions of well width for three distinct indium (In) fractions. Panels **a–c** display electron–electron and electron–hole-related transition energies between ISB and BTB, and panels **d–f** show the corresponding DME, while panels **g–i** depict the RLT associated with these transitions

changes. However, for BTB transitions—especially those involving excited electron states and ground-state holes—the modulation of the band structure by indium content becomes the dominant factor. These trends underscore the distinct physical mechanisms governing optical transitions in quantum-confined systems and highlight the need to consider both geometrical (well width) and compositional (In-fraction) factors when engineering optoelectronic properties. As the In-fraction increases, the band structure of the InGaN alloy is significantly altered, modifying the spatial overlap between electron and hole wave functions and thereby reducing the dipole matrix elements (DME). This reduction reflects a critical interplay between quantum confinement and material composition, indicating that the optical behavior of the quantum well system is governed by both geometrical dimensions and the electronic properties of the alloy. This relationship is central to understanding the trends in radiative lifetimes (RLT), as depicted in panels (g), (h), and (i) of Fig. 6, which offer deeper insight into how radiative recombination processes—specifically those associated with ISB and BTB transitions—respond to variations in well width and indium content. For ISB transitions involving the ground state electron, the radiative lifetime increases quasi-exponentially with well width. This trend is attributed to a reduction in quantum confinement and transition energy, both of which contribute to a decreased transition probability and hence a longer recombination time. Conversely, the RLT associated with BTB transitions between the first-excited electron state and the ground-state hole decreases with increasing well width. This reflects a growing overlap between the relevant wave functions, enhancing the DME and thus accelerating the recombination process. Interestingly, transitions involving the second-excited electron state reveal a more nuanced behavior: the radiative lifetime initially increases slightly and then decreases smoothly as the well widens. This dual trend emerges from the competing effects of decreasing transition energy and increasing DME. While a reduced transition energy tends to lengthen the radiative lifetime due to its inverse cubic dependence on energy (as outlined in Eq. 1), the simultaneous and dominant enhancement of the DME leads to a net decrease in lifetime for larger well widths—clearly visible in panel (h). Moreover, indium content has a substantial and distinct impact on the RLT depending on the type of transition. An increase in In-fraction leads to longer radiative lifetimes for ISB transitions. This is because, although the transition energy increases with In content, the DME decreases more significantly, and the combined effect results in a slower radiative process. In contrast, BTB transitions exhibit shorter radiative lifetimes with increasing In-fraction. This behavior arises from the bandgap narrowing and the resulting enhanced DME, which outweighs the reduction in transition energy and accelerates radiative recombination. These observations highlight the intricate balance between confinement geometry, material composition, and electronic structure in tailoring the optical performance of quantum well systems. This trend results from the simultaneous decrease in both the transition energy and the dipole matrix element, leading to an overall reduction in the radiative lifetime. Specifically, as shown in panel (g), for an indium fraction of $x = 0.1$, the radiative lifetime increases significantly—from 0.5 fs at a well thickness of $l = 0.2$ nm, to 1.8 fs at $l = 0.5$ nm, and further to 4.5 fs at $l = 1.0$ nm. This behavior can be explained by the reduced spatial confinement of charge carriers in wider wells, which decreases the overlap between electron and hole wave functions. The reduced recombination rate, manifested as an extended radiative lifetime, implies a lower emission efficiency per unit time but potentially enables higher total light output due to sustained carrier excitation. Such characteristics are particularly beneficial for high-efficiency light-emitting diodes (LEDs) and laser diodes employed in general lighting and display technologies, where stable and intense emission is critical [37]. In contrast, panel (h) reveals an opposite trend, where the radiative lifetime decreases as the well thickness increases across all indium compositions. For instance, at $x = 0.1$, the radiative lifetime reduces from 0.07 ns at $l = 0.2$ nm to 0.04 ns at $l = 0.5$ nm, and further to 0.02 ns at $l = 1.0$ nm. This behavior indicates the presence of increased non-radiative recombination processes and possibly stronger carrier scattering in wider wells. Notably, faster recombination dynamics observed for higher indium contents—such as $x = 0.3$, where radiative lifetimes decrease from 0.05 to 0.01 ns over the same well thickness range—are advantageous for high-speed optoelectronic applications, where rapid carrier recombination is essential [38]. These devices demand rapid response times, making them well-suited for telecommunications, high-frequency signal modulation, and data transmission systems. Additionally, panel (i) shows a near-saturation behavior in radiative lifetime across all well thicknesses. For $x = 0.1$, the radiative lifetime remains relatively stable, changing only slightly from 0.06 ns at $l = 0.2$ nm to 0.08 ns at $l = 1.0$ nm. Similar trends appear for $x = 0.2$ and $x = 0.3$, with lifetimes ranging roughly between 0.07 and 0.08 ns. This stability suggests that beyond a certain well thickness, the effects of reduced quantum confinement are offset by other factors such as defect-related recombination or carrier scattering. The resulting uniform radiative lifetime is beneficial for applications requiring consistent light emission, including solid-state lighting and optical sensing technologies [39]. The ability to maintain steady output despite minor variations in structural parameters ensures device reliability and consistent performance stability, which is crucial for practical applications.

Figure 7 illustrates the dependence of intra-subband (ISB) and band-to-band (BTB) optical transitions, dipole matrix elements (DME), and radiative lifetimes (RLT) of electrons on quantum well width at three distinct temperatures. Panels (a), (b), and (c) present the transition energies for electron–electron and electron–hole transitions. As the well width increases, the transition energies decrease, reflecting reduced quantum confinement in wider wells. This leads to a convergence of energy levels, especially pronounced at lower temperatures where thermal broadening is minimal, resulting in sharper level separations. Conversely, at higher temperatures, thermal excitation broadens energy states, causing increased overlap and diminished confinement effects. Panels (d), (e), and (f) show the corresponding DMEs for these transitions. The DME generally increases with well width, indicating stronger optical transitions due to greater spatial extension and overlap of electron wave functions. Temperature significantly modulates this trend: at lower temperatures, wave functions remain more confined with reduced overlap, whereas at elevated temperatures thermal excitation causes wave functions to delocalize, enhancing overlap and increasing DMEs—an effect particularly notable in holes owing to their larger effective mass and complex valence band coupling. Panels (g), (h), and (i) depict the radiative lifetimes associated with the optical transitions. In panel (g), RLT increases with well width, a trend accentuated at higher temperatures due to reduced spatial overlap from thermal broadening, which lowers radiative recombination probabilities. Panel (h) shows a decreasing RLT with increasing well width, suggesting dominant non-radiative recombination or enhanced scattering in thicker wells, effects amplified by increased temperature and carrier interactions. Lastly, panel (i) reveals an initial increase in RLT followed by stabilization as the well widens, indicating a nuanced interplay between quantum confinement, recombination dynamics, and defect-related processes. Temperature modulates this balance by affecting wave function distribution and confinement, thus influencing recombination pathways.

The radiative lifetime (RLT) values presented in panels (g), (h), and (i) provide critical insights into electron–electron and electron–hole interactions within quantum

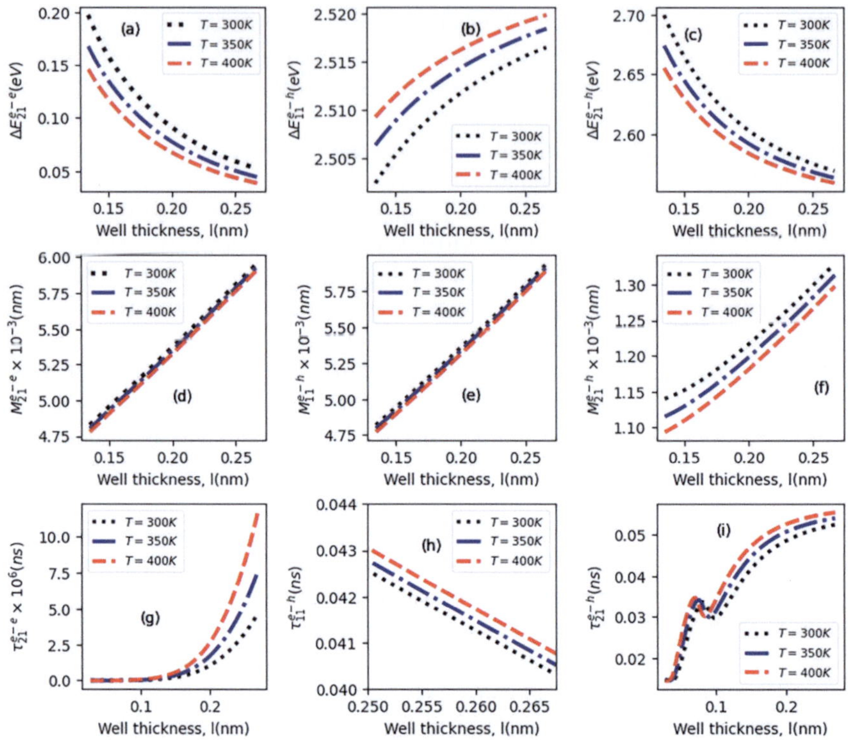

Fig. 7 Illustration of intra-subband and band-to-band optical energy transitions (ISB/BTB), dipole matrix elements (DME), and radiative lifetimes (RLT) of electrons as functions of well width for three distinct temperature values (T). Panels **a–c** display electron–electron and electron–hole-related transition energies between ISB and BTB, and panels **d–f** show the corresponding DME, while panels **g–i** depict the RLT associated with these transitions

wells, highlighting their dependence on well thickness and temperature. Expressed in femtosecond (fs) and nanosecond (ns) timescales, these lifetimes reflect the combined effects of quantum confinement, wave function overlap, and thermal excitation on radiative recombination processes—parameters with direct implications for the performance of optoelectronic devices. Panel (g) details radiative lifetimes associated with electron–electron transitions. For a narrow well width of 0.2 nm at 300 K, the RLT is approximately 1 fs, indicative of strong quantum confinement where close spatial proximity of electron wave functions enhances overlap and radiative recombination probability. At elevated temperatures, such as 400 K, and wider wells (e.g., 3 nm), the RLT increases to over 10 fs, consistent with literature [34]. This increase arises from thermal excitation reducing spatial overlap by weakening confinement, thereby lowering the recombination rate. In panel (h), the RLT trends for electron–hole transitions further illustrate the role of well width and temperature. At a narrow well width of 0.265 nm and 300 K, the RLT is about 0.040 ns, reflecting efficient radiative recombination due to strong confinement and significant wave function overlap. Such rapid recombination is highly desirable for devices requiring fast response times, including high-speed modulators and laser diodes [40]. Conversely, at 400 K with a thinner well width of 0.25 nm, the radiative lifetime decreases slightly to approximately 0.043 ns, indicating an acceleration of recombination dynamics. This behavior is driven by thermal effects that enhance non-radiative recombination pathways and carrier scattering mechanisms, thereby shortening the overall lifetime. Panel (i) illustrates the radiative lifetime trends for electron–hole transitions, emphasizing the interplay between quantum confinement and other recombination processes. At 300 K and a narrow well width of 0.1 nm, the radiative lifetime is about 0.03 ns, reflecting strong confinement and efficient radiative recombination. As the well width increases to 0.2 nm and temperature rises to 400 K, the radiative lifetime stabilizes near 0.05 ns, suggesting a balance between reduced quantum confinement and compensatory factors such as defect-related recombination or increased scattering at elevated temperatures [41]. These insights are vital for optimizing light-emitting diodes and laser diodes, as managing radiative lifetimes is key to balancing efficiency and stability [42]. The calculated capture time of approximately 10 fs in the InGaN quantum well with a 0.3 nm thickness at 400 K aligns well with reported trends in the literature, where capture times shorten with increasing confinement and reduced dimensionality. For example, experimental studies show capture times ranging from 500 fs in 20 nm quantum well interfaces (QWIs) to 60 fs in 30 nm quantum dots (QDs). Our result represents an extreme confinement regime, where enhanced carrier localization and elevated scattering rates at higher temperatures naturally yield shorter capture times [43]. A deeper understanding of these variations is critical for designing optoelectronic devices requiring rapid electron–hole recombination to support high-frequency operation. Stabilizing these dynamics is equally important for applications demanding consistent optical output, such as solid-state lighting and optical sensing, where device reliability under varying conditions is paramount. These findings underscore the necessity for precise control over quantum well dimensions and thermal management in advanced optoelectronic device engineering. Recommendations for future work include the precise control of quantum well thickness and indium fraction to fine-tune electronic states and optimize optical transitions for specific device functionalities, alongside the development of advanced temperature management strategies to mitigate thermal broadening effects and maintain desirable radiative recombination rates under varying thermal conditions. Additionally, incorporating non-radiative recombination pathways and scattering mechanisms into modeling frameworks will improve the prediction of real-world device behavior, while experimental validation and detailed material characterization are essential to ensure that theoretical insights effectively translate into practical device fabrication and performance. By integrating these approaches, future research can substantially enhance the design, efficiency, and reliability of InGaN-based optoelectronic devices, driving progress in solid-state lighting, high-speed communication systems, and optical sensing applications.

3 Conclusion

This chapter has systematically explored the intricate interplay between quantum well structural parameters, material composition, and temperature on the optical and electronic properties of InGaN quantum wells. Through detailed analysis of intra-subband (ISB) and band-to-band (BTB) transitions, we have demonstrated how variations in quantum well width, indium fraction, and thermal effects critically influence transition energies, dipole matrix elements (DME), and radiative lifetimes (RLT). The results highlight that quantum confinement and material composition govern electron and hole energy states differently, with indium content exerting a stronger influence on electron energies, while temperature variations predominantly affect hole energies. Key findings include the pronounced sensitivity of ISB transition energies to well width and indium composition, contrasted with the dominant role of the bandgap in BTB transitions. The quasi-linear increase of DME with well width and its decrease with rising indium content elucidate how wave function overlap and band structure modulations impact transition probabilities. Radiative lifetimes exhibit complex dependencies: increasing with well width for certain transitions due to reduced overlap, yet decreasing in others as enhanced scattering and non-radiative pathways emerge, especially under elevated temperatures. These insights collectively emphasize the necessity for a balanced optimization of quantum well geometry, alloy composition, and thermal management to tailor device performance for specific applications—ranging from high-efficiency LEDs requiring longer radiative lifetimes and stable emission, to high-speed modulators and lasers benefiting from rapid recombination dynamics.

Financial Funding Not applicable.

Data Availability Statement Not applicable.

Conflict of Interests The authors declare that they have no known competing financial interests or personal relationships that could have appeared to influence the work reported in this paper.

Institutional Review Board Statement The content of the article is original, has not been published elsewhere, and does not involve human or animal experiments.

References

1. M. Nami et al., Carrier dynamics and electro-optical characterization of high-performance GaN/InGaN core-shell nanowire light-emitting diodes. Sci. Rep. **8**(1), 501 (2018). https://doi.org/10.1038/s41598-017-18833-6
2. M.V. Maximov et al., Light emitting devices based on quantum well-dots. Appl. Sci. **10**(3), 3 (2020). https://doi.org/10.3390/app10031038
3. Y.B. Band, Y. Avishai, Low-dimensional quantum systems, chap. 13, in *Quantum Mechanics with Applications to Nanotechnology and Information Science*, ed. by Y.B. Band, Y. Avishai

(Academic Press, Amsterdam, 2013), pp. 749–823. https://doi.org/10.1016/B978-0-444-53786-7.00013-7

4. P.G. Eliseev, Radiative processes in InGaN quantum wells
5. E.A. Evropeitsev et al., State-of-the-art and prospects for intense red radiation from core–shell InGaN/GaN nanorods. Sci. Rep. **10**(1), 19048 (2020). https://doi.org/10.1038/s41598-020-76042-0
6. B. Li et al., Efficient and stable near-infrared InAs quantum dot light-emitting diodes. Nat. Commun. **16**(1), 2450 (2025). https://doi.org/10.1038/s41467-025-57746-1
7. M. Sheen et al., Highly efficient blue InGaN nanoscale light-emitting diodes. Nature **608**(7921), 56–61 (2022). https://doi.org/10.1038/s41586-022-04933-5
8. W. Gu et al., Highly coupled dynamically modulated electrocatalysts on wafer-scale InGaN/GaN nanowires on silicon for successive acidic photoelectrochemical water oxidation. Adv. Mater. 2501218. https://doi.org/10.1002/adma.202501218
9. B. Julsgaard, N. von den Driesch, P. Tidemand-Lichtenberg, C. Pedersen, Z. Ikonic, D. Buca, Carrier lifetime of GeSn measured by spectrally resolved picosecond photoluminescence spectroscopy. Photonics Res. **8**(6), 788–798 (2020). https://doi.org/10.1364/PRJ.385096
10. Y. Zhao, V. Wang, A. Javey, Molecular materials with short radiative lifetime for high-speed light-emitting devices. Matter **3**(6), 1832–1844 (2020). https://doi.org/10.1016/j.matt.2020.09.009
11. B. Lambert, B. Sermage, B. Deveaud, F. Clerot, A. Chomette, A. Regreny, Radiative and non-radiative recombination in GaAs-AlGaAs superlattices. Surf. Sci. **228**(1), 210–212 (1990). https://doi.org/10.1016/0039-6028(90)90293-H
12. C. Li, Z. Ji, J. Li, M. Xu, H. Xiao, X. Xu, Electroluminescence properties of InGaN/GaN multiple quantum well-based LEDs with different indium contents and different well widths. Sci. Rep. **7**(1), 15301 (2017). https://doi.org/10.1038/s41598-017-15561-9
13. D.J. Binks, P. Dawson, R.A. Oliver, D.J. Wallis, Cubic GaN and InGaN/GaN quantum wells. Appl. Phys. Rev. **9**(4), 041309 (2022). https://doi.org/10.1063/5.0097558
14. R. En-nadir, H. El-ghazi, Theoretical study of ISB conduction optical absorption and impurity binding energy associated with lowest excited states in QW with a new modulated potential. J. Theor. Appl. Phys. (2023). https://doi.org/10.30495/jtap.172317
15. R. En-nadir et al., Effects of laser and electric fields on ultrathin InGaN/GaN heterostructure: exploring binding energy, electron distribution, and diamagnetic susceptibility with harmonic and inharmonic potential. Phys. E Low-Dimens. Syst. Nanostruct. 115933 (2024). https://doi.org/10.1016/j.physe.2024.115933
16. R. En-nadir, M.A.B.-M. Kabatas, M. Tihtih, H.E. Ghazi, Linear and nonlinear optical absorption coefficients in InGaN/GaN quantum wells: interplay between intense laser field and higher-order anharmonic potentials. Heliyon (2023). https://doi.org/10.1016/j.heliyon.2023.e22867
17. R. En-nadir et al., Tailoring optoelectronic properties of InGaN-based quantum wells through electric field, indium content, and confinement shape: a theoretical investigation. Phys. B Condens. Matter **663**, 414976 (2023). https://doi.org/10.1016/j.physb.2023.414976
18. R. En-nadir, H. El Ghazi, A. Jorio, I. Zorkani, Inter and intra band impurity-related absorption in (In,Ga)N/GaN QW under composition, size and impurity effects. MATEC Web Conf. (2020)
19. R. En-nadir, H. El Ghazi, W. Belaid, A. Jorio, I. Zorkani, Intraconduction band-related optical absorption in coupled (In,Ga)N/GaN double parabolic quantum wells under temperature, coupling and composition effects. Results Opt. **5**, 100154 (2021). https://doi.org/10.1016/j.rio.2021.100154
20. R. En-nadir, H. El Ghazi, A. Jorio, I. Zorkani, H. Abboudi, F.A. Jabouti, Numerical study of temperature and electric field effects on the total optical absorption coefficient related-conduction-subband optical transitions in InGaN/GaN single parabolic QW. Fluid Dyn. Mater. Process. (2022)
21. R. En-nadir et al., Analyzing the combined influences of external electric field, impurity-location, in-content, and QW's number on donor-impurity binding energy in multiple quantum wells with finite squared potential. Opt. Quantum Electron. **55**(7), 597 (2023). https://doi.org/10.1007/s11082-023-04893-8

22. D. Feezell, S. Nakamura, Invention, development, and status of the blue light-emitting diode, the enabler of solid-state lighting. Comp. Rend. Phys. **19**(3), 113–133 (2018). https://doi.org/10.1016/j.crhy.2017.12.001
23. A. Caria et al., Quantum efficiency of InGaN–GaN multi-quantum well solar cells: experimental characterization and modeling. J. Appl. Phys. **131**(22), 224501 (2022). https://doi.org/10.1063/5.0076833
24. P. Ščajev et al., As-doped polycrystalline CdSeTe: localized defects, carrier mobility and lifetimes, and impact on high-efficiency solar cells. Adv. Energy Mater. **15**(10), 2403902 (2025). https://doi.org/10.1002/aenm.202403902
25. S. Chen et al., Carrier recombination suppression and transport enhancement enable high-performance self-powered broadband SbSe photodetectors. InfoMat **5**(4), e12400 (2023). https://doi.org/10.1002/inf2.12400
26. S.C. Lee, Y.F. Zhao, R.D. Schrimpf, M.A. Neifeld, K.F. Galloway, Comparison of lifetime and threshold current damage factors for multi-quantum-well (MQW) GaAs/GaAlAs laser diodes irradiated at different proton energies. IEEE Trans. Nucl. Sci. **46**(6), 1797–1803 (1999). https://doi.org/10.1109/23.819156
27. H. Zhao, G. Liu, J. Zhang, J.D. Poplawsky, V. Dierolf, N. Tansu, Approaches for high internal quantum efficiency green InGaN light-emitting diodes with large overlap quantum wells. Opt. Express **19**(104), A991–A1007 (2011). https://doi.org/10.1364/OE.19.00A991
28. R. En-nadir et al., Enhancing emission via radiative lifetime manipulation in ultrathin InGaN/GaN quantum wells: the effects of simultaneous electric and magnetic fields, thickness, and impurity. Nanomaterials **13**, 2817 (2023). https://doi.org/10.3390/nano13212817
29. S.M. Arif, A. Bera, A. Ghosh, M. Ghosh, Exploring noise-effect on the intraband transition lifetime of impurity doped quantum dots. Biointerface Res. Appl. Chem. **11**, 8639–8653 (2021)
30. D. Makhlouf, M. Choubani, F. Saidi, H. Maaref, Enhancement of transition lifetime, linear and nonlinear optical properties in laterally coupled lens-shaped quantum dots for tera-hertz range. Phys. E Low-Dimens. Syst. Nanostruct. **103**, 87–92 (2018)
31. A. Yariv, H.W. Moos, Quantum electronics. Am. J. Phys. **36**(2), 168–169 (1968)
32. F. Rajadell, J.I. Climente, J. Planelles, Excitons in core-only, core-shell and core-crown CdSe nanoplatelets: interplay between in-plane electron-hole correlation, spatial confinement, and dielectric confinement. Phys. Rev. B **96**(3), 035307 (2017). https://doi.org/10.1103/PhysRevB.96.035307
33. H. Gotoh, H. Ando, T. Takagahara, Radiative recombination lifetime of excitons in thin quantum boxes. J. Appl. Phys. **81**(4), 1785–1789 (1997). https://doi.org/10.1063/1.364034
34. Y. Arakawa, H. Sakaki, M. Nishioka, J. Yoshino, T. Kamiya, Recombination lifetime of carriers in GaAs-GaAlAs quantum wells near room temperature. Appl. Phys. Lett. **46**(5), 519–521 (1985). https://doi.org/10.1063/1.95578
35. J. Wilkes, E.A. Muljarov, Exciton effective mass enhancement in coupled quantum wells in electric and magnetic fields. New J. Phys. **18**(2), 023032 (2016). https://doi.org/10.1088/1367-2630/18/2/023032
36. W. Belaid, H. El Ghazi, I. Zorkani, A. Jorio, Pressure-related binding energy in (In,Ga)N/GaN double quantum wells under internal composition effects. Solid State Commun. **327**, 114193 (2021)
37. H. Xue et al., Recombination rate analysis of InGaN-based red-emitting light-emitting diodes. IEEE J. Quantum Electron. **59**(2), 1–9 (2023). https://doi.org/10.1109/JQE.2023.3246981
38. M. Palummo, M. Bernardi, J.C. Grossman, Exciton radiative lifetimes in two-dimensional transition metal dichalcogenides. Nano Lett. **15**(5), 2794–2800 (2015). https://doi.org/10.1021/nl503799t
39. D. Prasanth, D.V. Sunitha, P.R. Kumar, G.P. Darshan, Design strategies, luminescence mechanisms, and solid-state lighting applications of lanthanide-doped phosphorescent materials. ChemPhysMater (2024). https://doi.org/10.1016/j.chphma.2024.09.004
40. T.M. Ribeiro et al., Observation of thickness-independent ultrafast relaxation times in MPCVD few-layer graphene. Carbon **232**, 119700 (2025). https://doi.org/10.1016/j.carbon.2024.119700

41. Y. Hu, J. Xu, X. Ruan, H. Bao, Defect scattering can lead to enhanced phonon transport at nanoscale. Nat. Commun. **15**(1), 3304 (2024). https://doi.org/10.1038/s41467-024-47716-4
42. X. Mi, B.A. Rungo, X. Dong, H. Liu, X. Li, S. Wang, Enhanced efficiency and stability of organic light-emitting diodes via binary self-assembled monolayers of aromatic and aliphatic compounds on indium tin oxide. Org. Electron. **84**, 105752 (2020). https://doi.org/10.1016/j.orgel.2020.105752
43. T. Aggarwal et al., Investigation of ultrafast carrier dynamics in InGaN/GaN-based nanostructures using femtosecond pump-probe absorption spectroscopy. Phys. Status Solidi B **258**(10), 2100223 (2021). https://doi.org/10.1002/pssb.202100223

Optical Absorption and Oscillator Strength Modulation in GaAs-Based Quantum Dots

A. Fakkahi, M. Kirak, M. Jaouane, and A. Sali

1 Introduction

It is widely acknowledged that the use of low-dimensional structures in optical devices, sensors [1], and optoelectronics has received much attention in recent years. QDs, QWs, QWWs, and multilayer quantum dots (MQDs) are examples of low-dimensional structures. With technological progress, it has become possible to prepare QDs and MQDs [1–3]. The determination of the optical characteristics of QDs and MQDs has attracted considerable interest from researchers. The information about the optical characteristics of QDs and MQDs can be employed in designing optoelectronic devices [1–4].

It is known that the multilayer structures can confine the electron in several regions. The electron can experience a different confining potential in each region [5–8]. The use of MQDs in optoelectronic devices like light-emitting diodes and optical amplifiers has increased in recent years because these structures give better experimental results. The theoretical prediction of optical characteristics of MQDs is important because these structures are vital in fabricating novel optical devices with high efficiency [7, 9–11].

It is to be noted that many external factors like temperature, pressure, impurity, etc. can change the optical characteristics of QDs and MQDs. As a result, numerous researchers have contributed to this field of study. For instance, Jaouane et al. [12] studied the impacts of *P* and *T* on the PCS in MQD of the cylindrical shape. Fakkahi et al. [13] studied the optical rectification in a MSQD using the FEM. To get more information about MQDs, the reader can refer to [7–10, 12, 14–16].

A. Fakkahi (✉) · M. Jaouane · A. Sali
Laboratory of Solid-State Physics, Department of Physics, Faculty of Science Dhar El Mahraz, Sidi Mohamed Ben Abdellah University, Fez, Morocco
e-mail: abdelghani.fakkahi@usmba.ac.ma

M. Kirak
Department of Mathematics and Science Education, Faculty of Education, Yozgat Bozok University, Yozgat, Turkey

R. En-nadir (ed.), *Quantum Systems at the Nanoscale*, Lecture Notes in Nanoscale Science and Technology 45, https://doi.org/10.1007/978-3-032-06645-9_10

To study the physical characteristics of QDs and MQDs theoretically, we need information about the confining potential. Hitherto, many confining potentials have been used in the literature. Examples of the potentials are the Kratzer, Morse, Deng–Fan–Eckart, Woods–Saxon (WS), etc. [17–19]. The calculations of the physical characteristics of QDs and MQDs considering different confining potentials have been attended by many scientists [10, 12, 14, 15, 20–34]. In certain cases, the Schrödinger equation does not have an analytical solution for certain potentials. In this case, procedures based on the perturbation theory, diagonalization, factorization, and semi-linearization are used.

A numerical technique for resolving the Schrödinger equation (SE) is the finite Element Method (FEM). The FEM has very wide applications in physics, optics, electrochemistry, and plasma. In this method, the space is divided into a large number of meshes, and each mesh is considered as a partial element. Then, the desired differential equations are generalized to these components, and the variables will be calculated in the whole space [35–38]. Using the FEM, the SE is solved for a MSQD within the EMA. Then, eigenvalues and eigenstates are determined, thereby the absorption coefficients (ACs) are obtained.

Lorem ipsum dolor sit amet, consectetur adipiscing elit, sed do eiusmod tempor incididunt ut labore et dolore magna aliqua. Ut enim ad minim veniam, quis nostrud exercitation ullamco laboris nisi ut aliquip ex ea commodo consequat.

2 Theory and Calculation

We examine the confinement of an electron within a MSQD composed of $GaAs/AlGaAs/GaAs/AlGaAs$ layers, with a radius denoted as R_1. In the presence of a uniform static EF along the z-direction, we utilize the EMA to express the Hamiltonian of the considered system can be written in spherical coordinate systems as stated below:

$$\left[\frac{1}{m^*}\left(-\frac{\partial^2}{\partial r^2}+\frac{2}{r}\frac{\partial}{\partial r}+\frac{1}{r^2}\frac{\partial^2}{\partial \theta^2}\frac{\cot\theta}{r^2}\frac{\partial}{\partial\theta}+\frac{m^2}{r^2\sin^2\theta}\right)\right.$$
$$\left.+\,\eta r\cos\theta+V(r)-\frac{2Z}{\varepsilon\sqrt{r^2+d^2-2rd\cos\theta}}\right]\psi(r,\theta)=E\psi(r,\theta), \quad (1)$$

Here, m^* represents the effective mass of the electron, ε denotes the dielectric constant, η signifies the dimensionless measure of the EF, Z represents the charge of the impurity, and r and d correspond to the position coordinates of the electron and impurity, respectively.

The effective masses (dielectric constants) of materials are given by

$$m^*(\varepsilon)=\begin{cases} m_1^*(\varepsilon_1) & \text{for } \; 0\le r\le R_1 \;\text{ and }\; R_2\le r\le R_3 \\ m_2^*(\varepsilon_2) & \text{for } \; R_1<r<R_2 \;\text{ and }\; r>R_3 \end{cases} \quad (2)$$

where R_1 is the radius of core, R_2 and R_3 are inner and outer shell regions, respectively. $m_1^*(\varepsilon_1)$ and $m_2^*(\varepsilon_2)$ are the electron effective masses (dielectric constants) in $GaAs$ dot and $AlGaAs$ barrier materials, respectively.

The confining potential, denoted as $V(r)$, is defined as the sum of two components: the Kratzer–Fues potential (V_K) and the rectangular potential (V_i). The mathematical expression of the V_K is as follows:

$$V_K = -D\left(\frac{2R_0}{r} - \frac{R_0^2}{r^2}\right) \tag{3}$$

The rectangular potential V_i.

$$V_i = \begin{cases} 0 & \text{for} \quad 0 \le r \le R_1 \quad \text{and} \quad R_2 \le r \le R_3 \\ V_0 & \text{for} \quad R_1 < r < R_2 \quad \text{and} \quad r > R_3 \end{cases} \tag{4}$$

The graphical representations of Kratzer–Fues potential profiles are presented for different parameter values (see Fig. 8 in Chap. 4), and $V_0 = 228$, meV is the confinement potential height, which is serves as a characteristic value for D. Figure 1 in Chap. 11 reveals the form of schematic representation of the studied structure. Investigating the optical nonlinearity of nanostructures is crucial for their potential applications in devices. The system is subjected to a spatial electromagnetic wave of frequency ω, and considering the selection rules $\Delta l = \pm 1$, the analytical form of the total absorption coefficient can be derived by:

$$\alpha_T(\omega, I) = \alpha_L^{(1)}(\omega) + \alpha_N^{(3)}(\omega, I) \tag{5}$$

The above expression comprises two terms, the first denoting the linear optical absorption coefficient and the second representing the nonlinear optical absorption coefficient.

$$\alpha_L^{(1)}(\omega) = \omega\sqrt{\frac{\mu}{\varepsilon_R}}\frac{\sigma_V e^2 \left|M_{fi}\right|^2 \hbar\Gamma_{fi}}{\left(\hbar\omega - \Delta E_{fi}\right)^2 + \left(\hbar\Gamma_{fi}\right)^2}, \tag{6}$$

and

$$\alpha_N^{(3)}(\omega, I) = \omega\sqrt{\frac{\mu}{\varepsilon_R}}\left(\frac{I}{2\varepsilon_0 n_r c}\right)\frac{\sigma_V \left|M_{fi}\right|^2 \hbar\Gamma_{fi}}{\left[\left(\hbar\omega - \Delta E_{fi}\right)^2 + \left(\hbar\Gamma_{fi}\right)^2\right]^2}\left\{4\left|M_{fi}\right|^2 - \frac{\left(M_{ff} - M_{ii}\right)^2\left[3E_{fi}^2 - 4\Delta E_{fi}\hbar\omega + \hbar^2(\omega^2 - \Gamma_{if}^2)\right]}{E_{fi}^2 + (\hbar\Gamma_{if})^2}\right\}. \tag{7}$$

In the given context, the symbols used have the following meanings: σ_V denotes the electron density, μ represents the magnetic permeability, ε_0 corresponds to the dielectric permittivity of vacuum, ε_R stands for the real part of the relative permittivity, ω represents the angular frequency of the incident photon, $\hbar\omega$ signifies the energy of the incident photon, n_r denotes the refractive index of the dot material, c represents the speed of light, e corresponds to the electron charge, I signifies the intensity of the incident light, E_i and E_f represent the initial and final state energies, respectively, ΔE_{fi} represents the energy difference between the final and initial states, Γ_{if} is the relaxation rate for states i and f, τ denotes the relaxation time, and M_{if} stands for the dipole transition matrix element between states i and f. In this study, we have considered polarized electromagnetic radiation in the z-direction. The oscillator strength, a dimensionless physical quantity, is defined as the probability of the system interacting with the radiation field during transitions between energy levels. This quantity is determined using Fermi's golden rule [39]:

$$O_{fi} = \frac{2m^*}{\hbar^2} \Delta E_{fi} |M_{fi}|^2 \tag{8}$$

Here, ΔE_{fi} represents the energy difference between the final and initial states, and $|M_{fi}|^2$ denotes the squared absolute value of the transition matrix element $\langle \psi_f | r\cos\theta | \psi_i \rangle$.

3 Outcomes and Explanation

3.1 *Impact of Impurity Location and Electric Field on the Optical ACs*

In this part, we will delve into the computational results obtained from our theoretical investigation of the $\alpha_N^{(1)}(\omega)$, $\alpha_N^{(3)}(\omega, I)$, and $\alpha_T(\omega, I)$ intersubband optical absorptions, as well as the Oscillator Strength (OS). These investigations are focused on an impurity trapped in a $GaAs/Ga_{1-x}Al_xAs$ MSQD. The structure, depicted in Fig. 1, is characterized by specific geometry sizes, which are described in detail. Additionally, we present a graphical depiction illustrating the potential profile of the system. The variables δ_2, δ_1, and δ_3 represent the shell thickness, core radius, and well thickness, respectively. The position of the donor impurity is denoted by $\vec{d}$.

3.2 *Influence of Impurity Location on the Optical ACs*

Examining the impact of impurities on the electronic states of nanostructured semiconductors is of utmost importance due to its potential to substantially modify the optoelectronic characteristics of the semiconductor configuration.

In Fig. 2, we generated plots of the $\alpha^{(1)}(\omega)$, $\alpha^{(3)}(\omega, I)$, and total absorption coefficients (ACs) as functions of $\hbar\omega$, while keeping the parameters fixed: $\xi = 0.0$, $\delta_1 = 0.5a, \delta_2 = 0.4a, \delta_3 = 0.4a^*$, and $I = 0.5\,\mathrm{MW/cm^2}$. Four cases are considered:

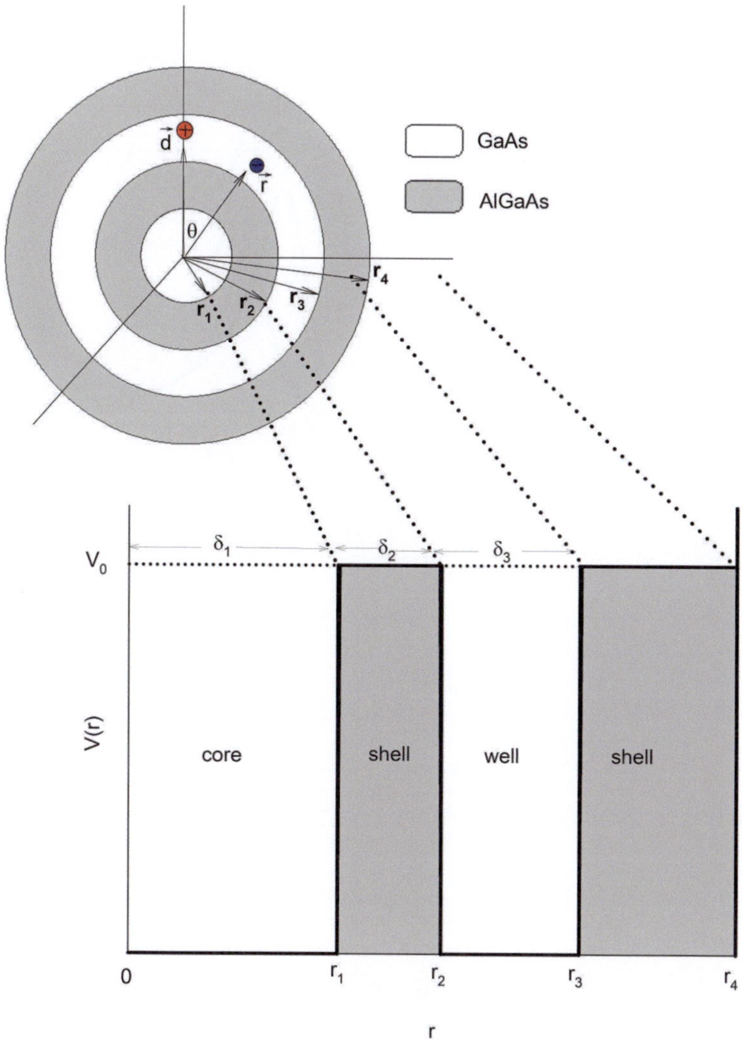

Fig. 1 The MSQD illustration and its rectangular potential profile

the absence of impurity ($Z = 0$, depicted by black lines), an impurity positioned at the center of the structure ($Z = 1, d = 0$, indicated by red lines), an impurity located at the right edge of the core ($Z = 1, d = \delta_1$, shown by blue lines), and an impurity situated at the left edge of the well ($Z = 1, d = \frac{\delta_2+\delta_3}{2}$, illustrated by green lines).

As depicted in the graph, across all scenarios, the $\alpha^{(1)}(\omega)$ increases, while the $\alpha^{(3)}(\omega, I)$ decreases initially, attains its peak value at approximately $\hbar\omega \approx \Delta E$, after which it begins to decline. This behavior is consistent with previous findings in the literature. Notably, when the impurity is placed at the middle of the MSQD ($Z = 1$,

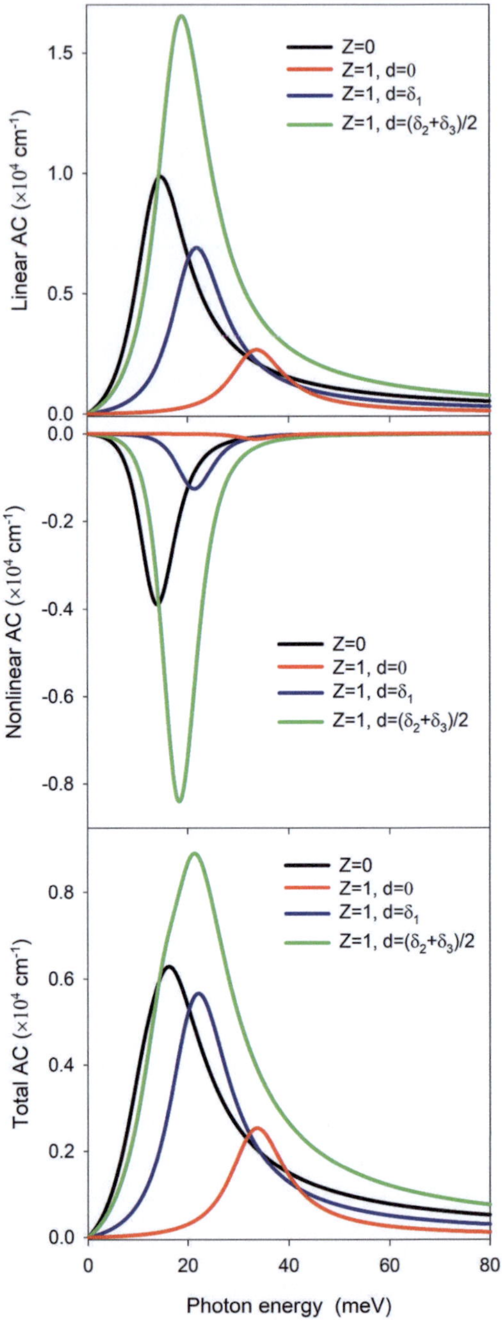

Fig. 2 The $\alpha_L^{(1)}(\omega)$, $\alpha_N^{(3)}(\omega, I)$ and $\alpha_T(\omega, I)$ versus the $\hbar\omega$ for different position of the impurity

$d = 0$), a noteworthy blue shift is observed in the resonance peaks of the optical ACs, accompanied by a diminution in their magnitudes. This blue-shift can be attributed to the increased resonance energy caused by the existence of the impurity, which leads to the formation of an intermediate band that resides between the valence and conduction bands of the dot material.

Due to the attractive Coulomb potential exerted by the impurity, the ground state ($1s$) level experiences a more pronounced downward shift in comparison to the excited state ($1p$), resulting in an augmentation of the resonance energy, as depicted in the left panels of Fig. 3. Additionally, the diminish in the peak amplitudes is associated with a diminution in the overlapping of the wave functions.

Figure 4a, b clearly shows that both wave functions associated with the $1s$ and $1p$ states are confined to the well region, indicating a greater level of overlap exists between the wave functions.

Conversely, in the scenario where $Z = 1$ and $d = 0$, the overlap between the wave functions diminishes as a result of the ground state wave function being localized in the core region and the excited state wave function being localized in the well region (Fig. 4c, d). Consequently, the peak intensity is reduced when the impurity is positioned at $d = 0$. However, when the impurity is shifted from the center of the MSQD ($d = 0$) towards the left edge of the well region ($d = \frac{\delta_2+\delta_3}{2}$), the resonance peak intensity increases and displaces to lower energy. This behavior can be attributed to the diminution in resonance energy values when the shallow donor defect is moved away from the heart of the MSQD, as observed in Fig. 3b, d, f. Additionally, the overlap of the wave functions becomes relatively high in the well because both wave functions are entirely bound within that zone (Fig. 4g, h). Figure 3 provides insight into the effects of δ_1, δ_2, and δ_3 on the peak value of the total AC and the resonance energy. The computations are carried out for four scenarios: absence of an impurity ($Z = 0$), and the presence of an impurity ($Z = 1$) positioned at $d = 0$, $d = \delta_1$, and $d = \frac{\delta_2+\delta_3}{2}$. Figure 3a displays the changes of the maximum point of the $\alpha_T(\omega, I)$ peak versus the δ_1. For small core radius (less than $0.6a^*$), the maximum point of the $\alpha_T(\omega, I)$ peak sharply diminishes with increasing $d = \delta_1$, while for core radius between $0.6a^*$ and $1a^*$, it grows, resulting in a shoulder. As the donor defect transitions from the center to the left edge of the well, the corresponding shoulder experiences a minor reduction. Additionally, in the case of a wide core radius, the presence and position of the impurity no longer significantly impact the peak value of the $\alpha_T(\omega, I)$. These findings align with the results documented in the in Ref. [40]. In Fig. 3b, the dependence of the resonance energy on the δ_1 is illustrated for the aforementioned scenarios. Across all cases, the resonance energy exhibits an initial increase, reaches a peak value, and then diminishes as the $d = \delta_1$ expands. When the impurity is shifted from the left edge of the well to the center of the MSQD, the amplitude of the resonance energy peak intensifies. Furthermore, the presence of the impurity leads to higher resonance energy compared to its absence, which can be attributed to the same underlying factors discussed earlier in Fig. 2.

In Fig. 3c, the relationship between the highest-point of $\alpha_T(\omega, I)$ and the δ_2 is depicted. Across all cases, the highest-point of $\alpha_T(\omega, I)$ experiences a substantial decrease as the shell thickness increases. This decrease occurs rapidly, eventually

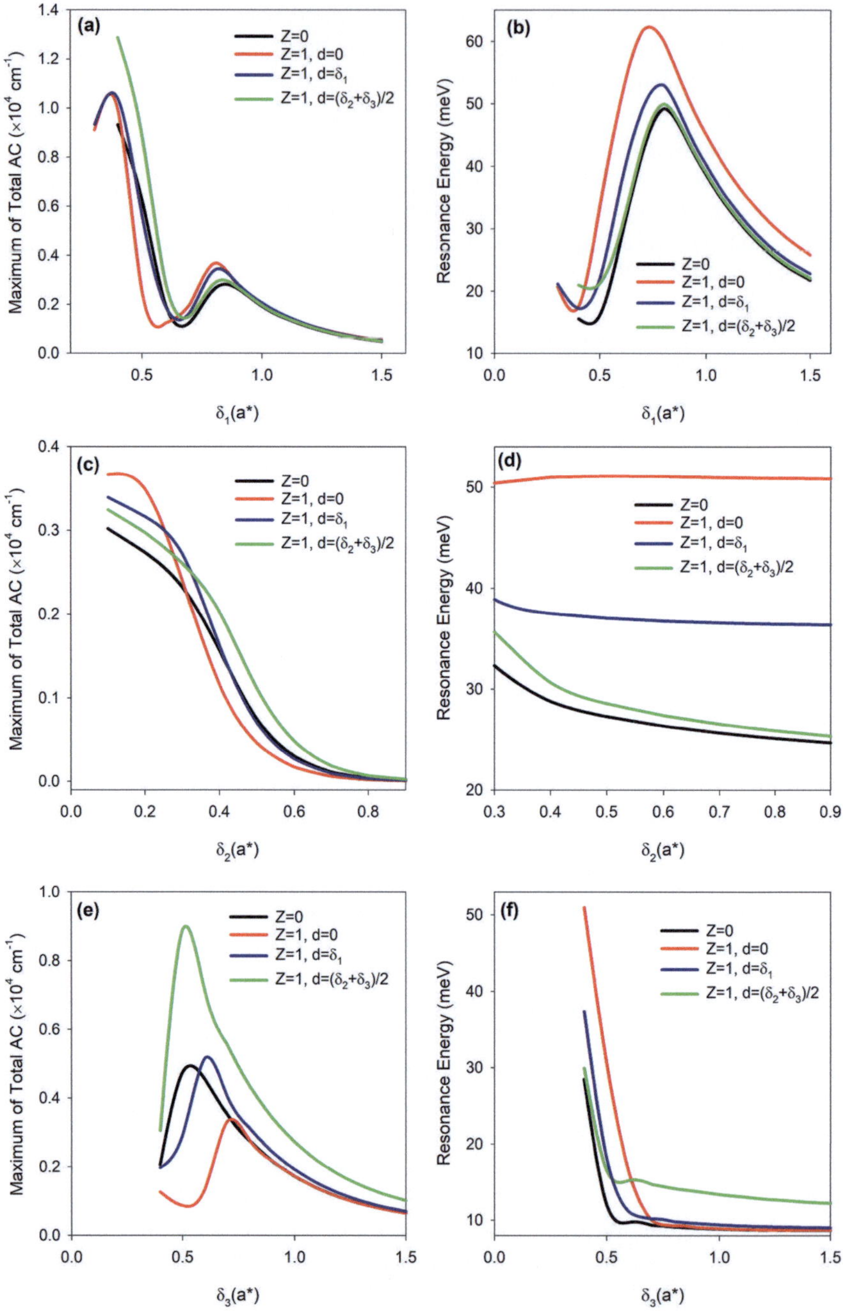

Fig. 3 The change in the peak value of the total AC (depicted in the left panel) and the resonance energy (shown in the right panel) is examined in relation to the δ_1 (**a**, **b**), δ_2 (**c**, **d**), and δ_3 (**e**, **f**) for various impurity locations

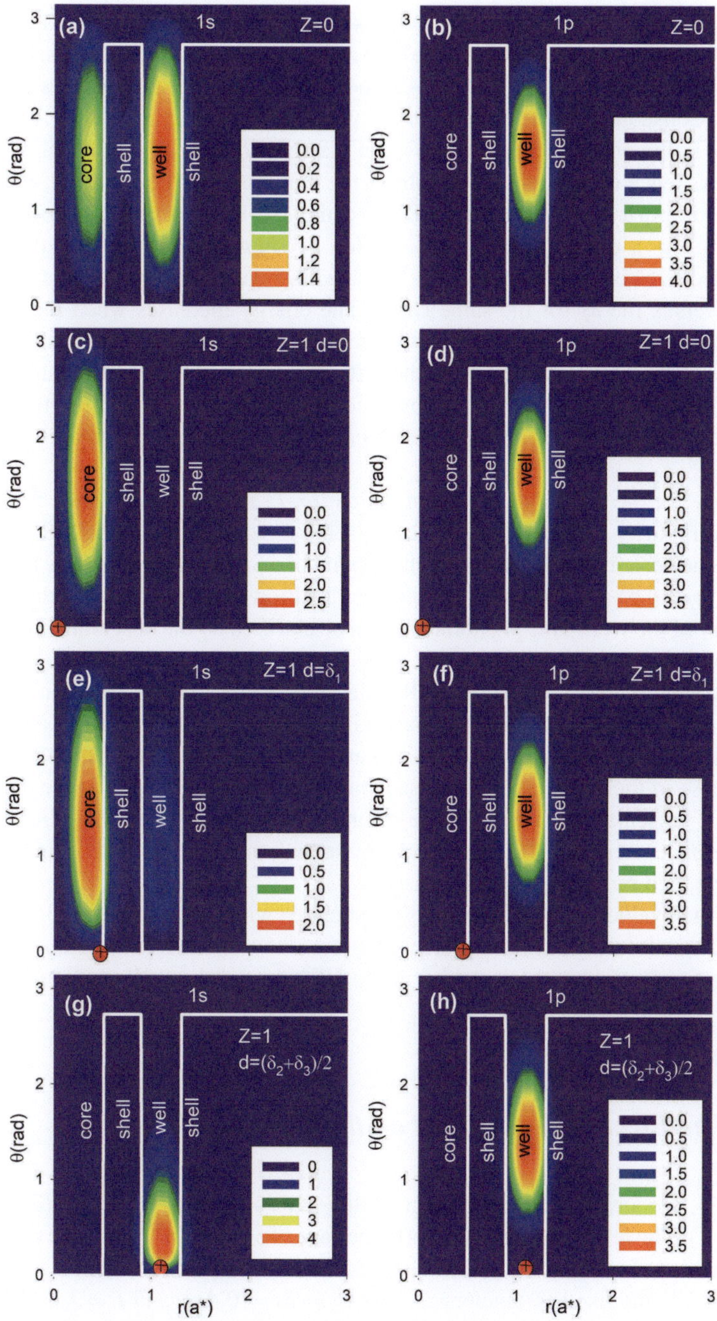

Fig. 4 The probability distribution of the $1s$ state (left panel) and $1p$ state (right panel) for several impurity position and constant parameters are as follows: $\delta_1 = 0.5a_d^*$, $\delta_2 = 0.4a_d^*$, $\delta_3 = 0.4a_d^*$, and $I = 1.0\,\text{MW/cm}^2$

approaching zero as the δ_2 continues to expand. This behavior can be attributed to the diminishing coupling between the core and the well as the δ_2 increases, leading to the confinement of the wave functions associated with the $1s$ and $1p$ states in distinct regions. Consequently, the overlap of the wave functions diminishes, resulting in a decrease in the electronic dipole transition matrix. Additionally, for small values of δ_2, the existence of the impurity amplifies the maximum of the $\alpha_T(\omega, I)$, especially when the impurity is situated at the MSQD center. As the δ_2 surpasses a critical value (above $0.35a^*$), the maximum of the $\alpha_T(\omega, I)$ augments as the ionized defect is shifted from the center to the left edge of the well. Figure 3d illustrates the resonance energy versus the δ_2 for the same cases. The energy increases when the defect is introduced and positioned at the MSQD center. Moreover, when the shallow donor defect is shifted from the center to the left edge of the well, the resonance energy reduces and approaches the values observed in the absence of the impurity ($Z = 0$), with the exception of the $Z = 1$, $d = 0$ case where the energy remains relatively constant.

In Fig. 3e, the impact of the δ_3 on the $\alpha_T(\omega, I)$ maximum is shown for several cases of impurity location. The existence of the ionized defect at the MSQD center reduces the peak magnitude of the highest-point of $\alpha_T(\omega, I)$. However, the maximum point of $\alpha_T(\omega, I)$ augments as the impurity transitions from the center to the leftmost boundary of the well. This phenomenon can be clarified by the observation that when the impurity is located at the center ($d = 0$), the $1s$ ($1p$) wave function becomes confined to the core (well) region. Resulting in lower overlap between the wave functions. In contrast, when the impurity is positioned at the leftmost boundary of the well ($d = \frac{\delta_2+\delta_3}{2}$), both wave functions are restricted to the well region, leading to an increase in their overlap and, consequently, a higher magnitude of the $\alpha_T(\omega, I)$. Figure 3f presents the changes of the resonance energy against the δ_3. In all cases, the resonance energy exhibits a sharp decrease with the well width, reaching a lowest value and then remaining almost constant.

For small values of the δ_3, the energy associated with the $Z = 1, d = 0$ scenario is elevated and decreases as the impurity progresses towards the well region. However, for larger well widths, the impurity at the left edge of the well provides a higher resonance energy value.

3.3 The Electric Field Effect on the Optical ACs

In Fig. 5, we investigate the impact of the EF on the $\alpha_L^{(1)}(\omega)$, $\alpha_N^{(3)}(\omega, I)$, and $\alpha_{(T)}(\omega, I)$. The left panel corresponds to the case of $\delta_1 = 0.6a^*$, while the right panel corresponds to $\delta_1 = 0.8a^*$. The calculations are performed for different EF values (ξ = 0, 1, and 2) and fixed parameters: $Z = 0$, $\delta_2 = \delta_3 = 0.4a^*$, and $I = 1.0\,\text{MW/cm}^2$.

For both cases of $\delta_1 = 0.6a^*$ and $\delta_1 = 0.8a^*$, a raise in the EF leads to a redshift of the resonant peaks of $\alpha_L^{(1)}(\omega)$, $\alpha_N^{(3)}(\omega, I)$, and $\alpha_{(T)}(\omega, I)$. This displacement is due to the reduction of the resonance energy caused by the applied EF, as demonstrated

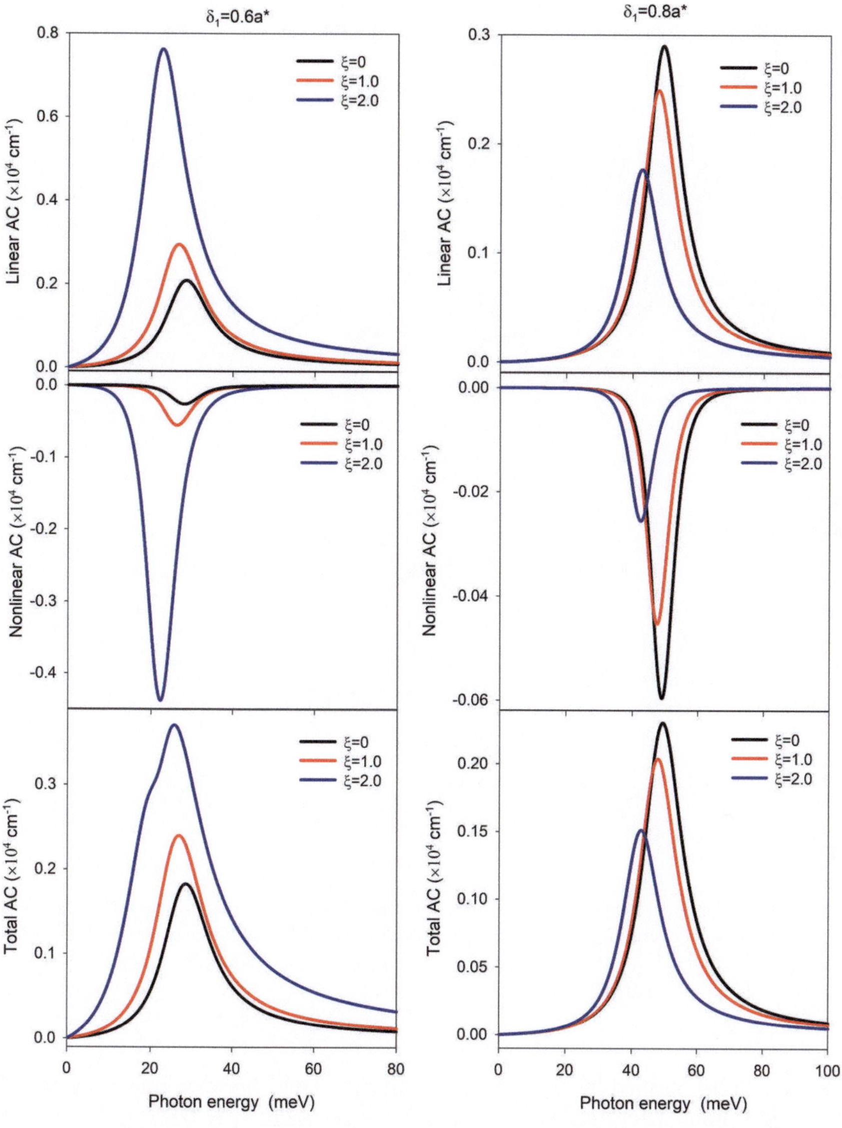

Fig. 5 The $\alpha_L^{(1)}(\omega)$, $\alpha_N^{(3)}(\omega, I)$ and $\alpha_T(\omega, I)$ versus the $\hbar\omega$ for several values of the EF, for two cases $\delta_1 = 0.6a_d^*$ (left panel) and $\delta_1 = 0.8a_d^*$ (right panel), with $\delta_2 = 0.4a_d^*$, $\delta_3 = 0.4a_d^*$, $Z = 0$ and $I = 1.0\,\text{MW/cm}^2$

in Fig. 3. In the case of $\delta_1 = 0.6a^*$, the magnitude of the resonant peaks of $\alpha_L^{(1)}(\omega)$, $\alpha_N^{(3)}(\omega, I)$, and $\alpha_{(T)}(\omega, I)$ increases as the EF strength increases. However, in the case of $\delta_1 = 0.8a^*$, the resonant peaks amplitude decreases with the EF. This phe-

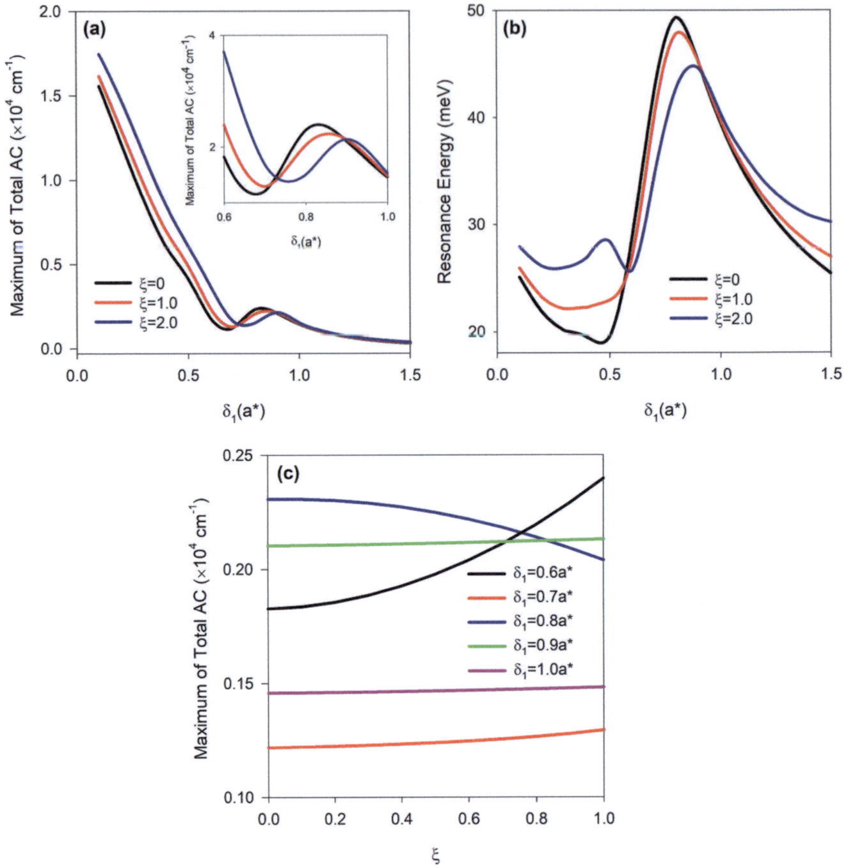

Fig. 6 The alteration of the peak value of $\alpha_T(\omega, I)$ and the resonance energy with respect to the δ_1 (**a**, **b**) is examined for various electric field values. Additionally, the maximum of $\alpha_T(\omega, I)$ is plotted against the external EF for different δ_1 values. The unchanged parameters are: $\delta_2 = 0.4a_d^* = \delta_3 = 0.4a_d^*$, $Z = 0$ and $I = 1.0\,\mathrm{MW/cm^2}$

nomena can be clarified by the overlapping of the $1s$ and $1p$ wave functions, which augments (diminishes) with the external EF in the case of $\delta_1 = 0.6a^*$ ($\delta_1 = 0.8a^*$).

In Fig. 6a, the maximum of the $\alpha_T(\omega, I)$ is plotted against the δ_1 for several values of the EF ($\xi = 0$, 1, and 2). The identical fixed parameters utilized in Fig. 5b are employed. The maximum of the $\alpha_T(\omega, I)$ reduces linearly with the δ_1 until it reaches a critical point of δ_1 ($0.7a^*$). After surpassing this crucial threshold, it begins to rise marginally and demonstrates a peak within the interval of $0.7a^*$ and $1a^*$, as depicted in the inset diagram. The optical ACs are proportional to the electron density, which is inversely proportional to the core volume ($1/V$). Therefore, the maximum of the $\alpha_T(\omega, I)$ decreases with the δ_1. However, there is a critical point where the behavior changes due to other factors such as the overlapping of wave functions and the influence of the EF, as discussed earlier. Consequently, the maximum values of $\alpha_T(\omega, I)$ obtained for smaller core radii are greater than those obtained for larger core radii. Furthermore, for core radius values (δ_1) smaller than $0.7a^*$, the electric field exhibits a significant impact, resulting in an increase in the maximum of $\alpha_T(\omega, I)$ with the EF. However, once the critical values of δ_1 ($0.7a^*$) are exceeded, the maximum of $\alpha_T(\omega, I)$ becomes less sensitive to the EF and displays an inverse behavior, diminishing with the EF, as demonstrated in the inset figure. Figure 6b displays the relationship between the resonance energy and the δ_1 for three different values of the EF ($\xi = 0$, 1, and 2). The resonance energy exhibits a complex behavior with respect to the augment in both the δ_1 and the EF. Our findings indicate that within the range of $0.6a^* \leq \delta_1 \leq 0.94a^*$, the resonance energy rapidly rises as the δ_1 expands, reaching a maximum value near $\delta_1 = 0.8a^*$. Furthermore, within this range, the resonance energy diminishes with an increase in the EF. However, for $\delta_1 \leq 0.6a^*$ and $0.9a^* \leq \delta_1$, it is evident that the resonance energy rises with the EF, particularly for $\delta_1 \leq 0.6a^*$ where the EF exhibits a significant influence. In Fig. 6c, we present the variation of the highest-point of $\alpha_T(\omega, I)$ versus the EF strength for different δ_1 values ($\delta_1 = 0.6a^*, 0.7a^*, 0.8a^*, 1a^*$). The figure clearly illustrates that the highest-point of $\alpha_T(\omega, I)$ is less responsive to changes in the EF for $\delta_1 = 0.7a^*, 0.9a^*$, and $1a^*$. However, the $\alpha_T(\omega, I)$ undergoes significant changes with the EF in the cases of $\delta_1 = 0.6a^*$ and $\delta_1 = 0.8a^*$, but in opposite directions. Specifically, it increases for $\delta_1 = 0.6a^*$ and decreases for $\delta_1 = 0.8a^*$. To further investigate the impact of these two δ_1 values on the optical ACs, we examined Fig. 5, and our findings align well with previous research. In Fig. 7, the ACs of the $1s - 1p$ transition are investigated against $\hbar\omega$ for different values of the EF ($\xi = 0$, 0.5, 1, and 1.5). The computations are carried out for two different values of the shell width, $\delta_2 = 0.4a_d^*$ and $\delta_2 = 0.6a_d^*$, with fixed parameters $\delta_1 = 0.6a_d^*$, $\delta_3 = 0.4a_d^*$, $Z = 0$, and $I = 2.0\,\text{MW/cm}^2$.

In both cases of $\delta_2 = 0.4a_d^*$ and $\delta_2 = 0.6a_d^*$, the resonant peaks of the optical ACs move toward lower energy (red shift) and their intensity enhances with the augmentation the EF. This behavior is attributed to the effect of the EF on the $1s$ and $1p$ electronic wave functions. The electric field causes the wave functions to become more confined in the same region, thereby reducing the resonance energy, as depicted in Fig. 8b.

This reduction in resonance energy causes a red shift in the peak position of the optical ACs. Furthermore, the rises in the EF strength enhances the resonant peaks intensity, indicating a stronger coupling between the $1s$ and $1p$ states under the influence of the electric field. Overall, the EF plays a crucial role in modifying the optical characteristics of the $1s - 1p$ transition, causing red shifts in the resonant peaks and amplifying their magnitudes.

In Fig. 8a, b, the $\alpha_T(\omega, I)$ peak maximum point and the resonance energy are plotted versus the δ_2 for several values of the EF. It can be observed that both the $\alpha_T(\omega, I)$ peak maximum point and the resonance energy diminishes with increasing δ_2. Furthermore, the effect of the EF on the $\alpha_T(\omega, I)$ peak maximum point and the resonance energy depends on the value of the δ_2. For small values of δ_2, the EF has no significant impact on the $\alpha_T(\omega, I)$ peak maximum point and the resonance

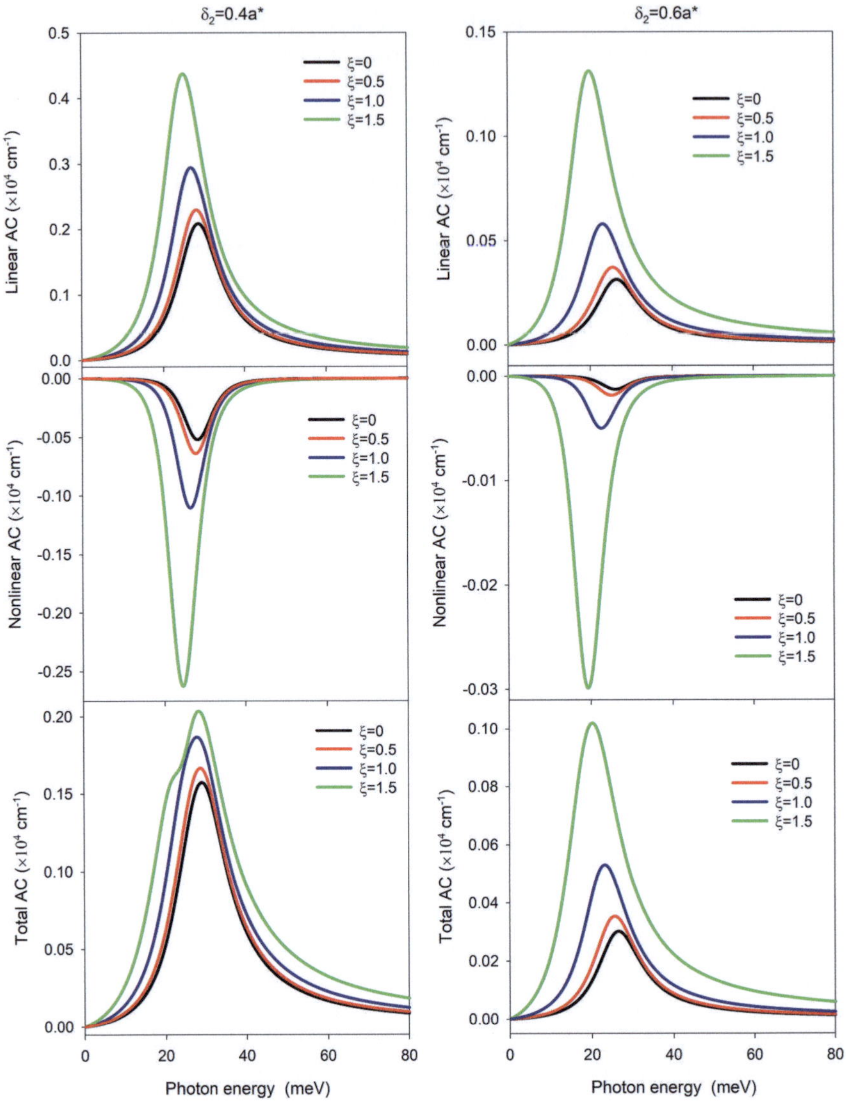

Fig. 7 The $\alpha_L^{(1)}(\omega)$, $\alpha_N^{(3)}(\omega, I)$ and $\alpha_T(\omega, I)$ against the $\hbar\omega$ for several values of the EF, for both cases $\delta_2 = 0.4a_d^*$ (left panel) and $\delta_2 = 0.6a_d^*$ (right panel), with $\delta_1 = 0.6a_d^*$, $\delta_3 = 0.4a_d^*$, $Z = 0$ and $I = 2.0\,\mathrm{MW/cm^2}$

energy. However, for larger values of δ_2, these quantities become more sensitive to changes in the EF. In particular, the $\alpha_T(\omega, I)$ peak maximum point rises with the EF, while the resonance energy reduces with the EF.

Moving to Fig. 9, the change of the optical ACs of the $1s - 1p$ transition is shown versus $\hbar\omega$ for different values of the EF (ξ = 0, 0.5, 1, and 1.5) and two different

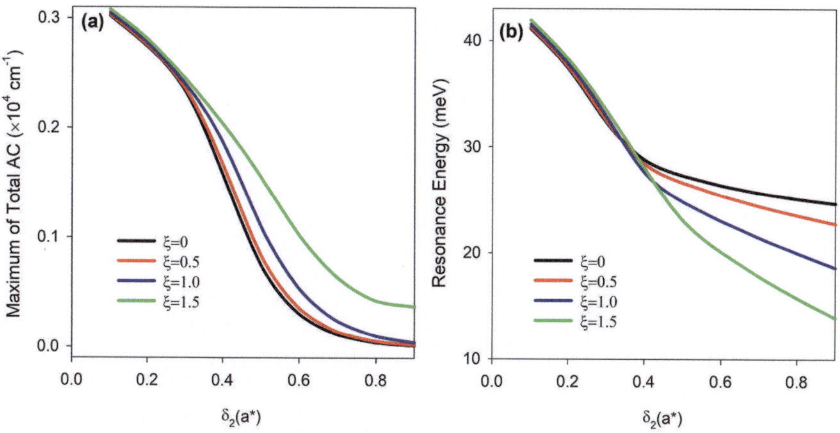

Fig. 8 The changes in the maximum value of $\alpha_T(\omega, I)$ (**a**) and the resonance energy (**b**) are examined with respect to the δ_2 for different values of the EF field, while keeping the other parameters fixed. At $\delta_1 = 0.6a_d^*$, $\delta_3 = 0.4a_d^*$, $Z = 0$ and $I = 2.0\,\text{MW/cm}^2$

values of the δ_3, $\delta_3 = 0.4a_d^*$ (left panel) and $\delta_3 = 0.7a_d^*$ (right panel). The δ_1, δ_2, and optical intensity are kept constant at $\delta_1 = 0.6a_d^*$, $\delta_2 = 0.4a_d^*$, and $I = 0.1\,\text{MW/cm}^2$, respectively.

The figure clearly demonstrates that the optical ACs' peak intensity exhibits a noticeable rise as the EF value increases for all scenarios. The EF enhances the overlapping of the wave functions involved in the $1s - 1p$ transition, resulting in a higher peak magnitude.

Furthermore, in the case where $\delta_3 = 0.7a_d^*$, the peak position of the ACs moves towards higher energy levels as the EF increases. On the other hand, for the case of $\delta_3 = 0.4a_d^*$, the peak position shows a slight displacement towards lower energy (red shift) as the EF increases. Moreover, it is worth noting that the values of the optical ACs obtained for the $\delta_3 = 0.7a_d^*$ case are higher than those obtained for the $\delta_3 = 0.4a_d^*$ case. This difference can be attributed to the variation in the well width, which affects the wave function distribution and, consequently, the optical ACs intensity. In summary, the EF enhances the peak intensity of the optical ACs and affects the peak position differently depending on the well width, resulting in either a red shift or a blue shift. Additionally, the values of the optical ACs are higher for larger values of the well width.

In Fig. 10a, b, the $\alpha_T(\omega, I)$ peak maximum point and the resonance energy are plotted against the δ_3 for various EF values. The fixed parameters used for these calculations are $Z = 0$, $\delta_1 = 0.6a_d^*$, $\delta_2 = 0.4a_d^*$, and $I = 0.1\,\text{MW/cm}^2$.

Figure 10a shows that the total optical AC exhibits a notable reliance on the EF, with a positive correlation between the EF and its magnitude across all well widths. This observation highlights the significant contribution of the electric field in augmenting the total optical AC, irrespective of the specific well width. On the other hand, Fig. 10b demonstrates that the presence of an EF has a considerable

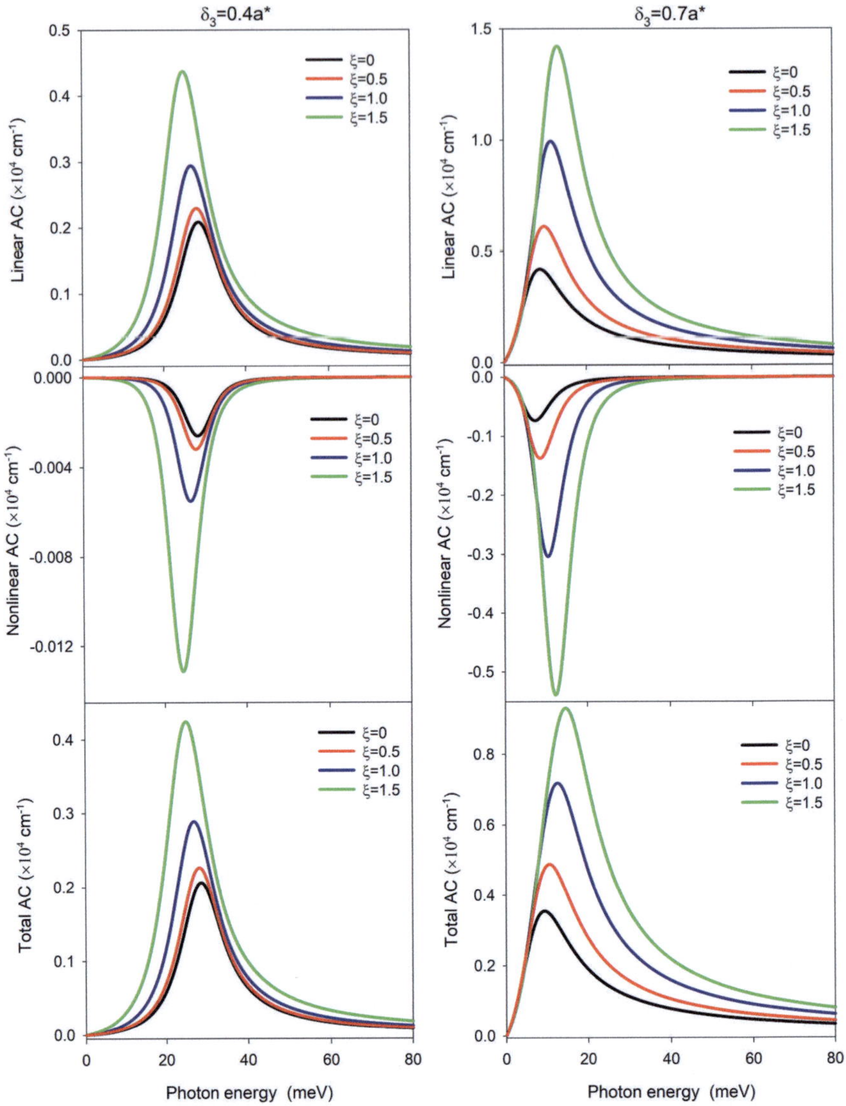

Fig. 9 The $\alpha_L^{(1)}(\omega)$, $\alpha_N^{(3)}(\omega, I)$ and $\alpha_T(\omega, I)$ against the $\hbar\omega$ for several values of the EF, for two cases $\delta_3 = 0.4a_d^*$ (left panel) and $\delta_3 = 0.7a_d^*$ (right panel), with $\delta_1 = 0.6a_d^*$, $\delta_2 = 0.4a_d^*$, $Z = 0$ and $I = 0.1\,\text{MW/cm}^2$

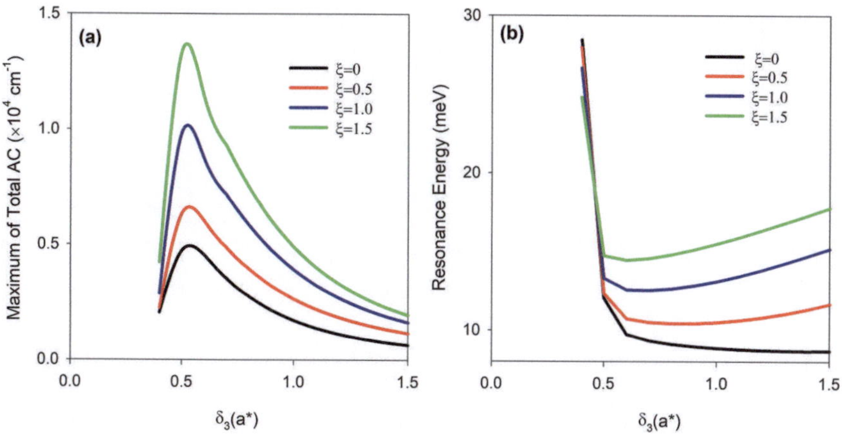

Fig. 10 The changes in the maximum value of $\alpha_T(\omega, I)$ (**a**) and the resonance energy (**b**) are examined with respect to the δ_3 for various values of the EF, while keeping the other factors fixed: $\delta_1 = 0.6a_d^*$, $\delta_2 = 0.4a_d^*$, $Z = 0$ and $I = 0.1\,\text{MW/cm}^2$

effect on the resonance energy for larger values of the δ_3 (greater than $0.5a_d^*$). In this case, the resonance energy augments with the EF. Nevertheless, when considering narrow well widths, the resonance energy remains unaffected by changes in EF. These results suggest that the influence of the EF on the total optical AC and the resonance energy depends on the well thickness. For larger well widths, the EF enhances both quantities. However, for smaller well widths, the electric field does not affect the resonance energy while still significantly impacting the total optical AC.

Overall, these findings highlight the complex interplay between the electric field, well width, and optical properties, providing valuable insights into the behavior of the system under different experimental conditions.

3.4 The Impact of the Electric Field and Impurity Location on the Oscillator Strength

In Fig. 11a, the changes of the oscillator strength (OS) is shown against the δ_1 for several values of the EF. The constant parameters used for these calculations are $Z = 0$, $\delta_2 = 0.4a_d^*$, and $\delta_3 = 0.4a_d^*$. From the figure, it can be observed that the OS exhibits a complicated behavior depending on the δ_1. The OS is strongly dependent on the δ_1, and it shows lower values when the δ_1 is between $0.5a_d^*$ and $1a_d^*$. However, it is interesting to note that the OS is not significantly affected by the EF. The EF does not cause a significant change in the OS for most δ_1 values. This behavior is consistent with previous studies, such as the one referenced as Ref. [40]. Figure 11b shows the changes of the OS with the EF strength for several values of the δ_1, using

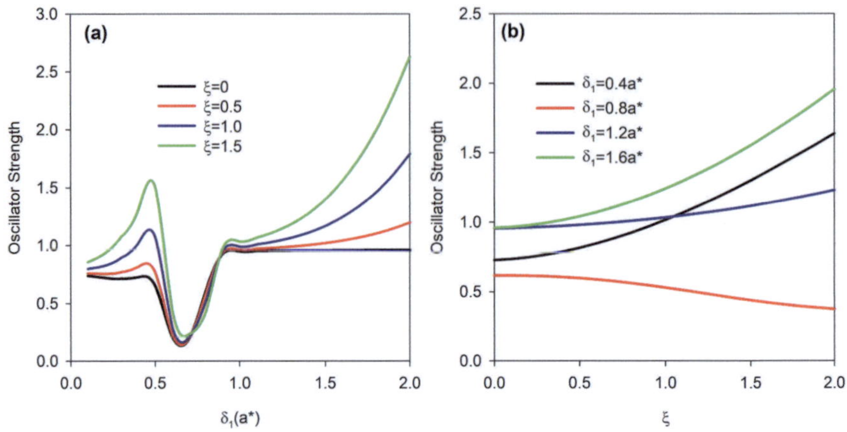

Fig. 11 **a** The OS against the δ_1 for several values of the EF. **b** The OS against the EF for several values of the δ_1 and for constant factors $\delta_2 = 0.4a_d^*$, $\delta_3 = 0.4a_d^*$, and $Z = 0$

the same variables as in Fig. 11a. In this case, it can be observed that the OS generally augments with the EF, but for the case of $\delta_1 = 0.8a_d^*$ where it exhibits an opposite behavior and decreases with the EF. These findings indicate that the OS is primarily influenced by the core radius, showing a complex behavior with lower values in the range of $0.5a_d^*$ to $1a_d^*$. The EF, on the other hand, has a more nuanced effect on the OS, generally increasing it except for specific cases such as $\delta_1 = 0.8a_d^*$ where it leads to a decrease. These results provide insights into the relationship between the structural parameters, EF, and OS, contributing to a better understanding of the optical characteristics of the system. In Fig. 12a, b, the oscillator strength (OS) and dipole matrix elements (DME) are plotted versus the δ_2 for several values of the EF (ξ = 0, 0.5, 1 and 1.5). It can be observed that both the OS and DME display analogous behavior: they diminish with increasing δ_2 and increase with higher EF intensities. This behavior can be described in the manner follows: as the δ_2 augments, there is a decline in the coupling between central area and the well. Consequently, the $1s$ and $1p$ wave functions become more localized to their respective areas, resulting in diminished overlap between the wave functions. This reduced overlap leads to a decrease in the OS and DME as the δ_2 increases.

Furthermore, the variation of the OS against impurity location is studied for different values of the δ_1 (Fig. 13a), δ_2 (Fig. 13b), and δ_3 (Fig. 13c). The fixed parameters used for these calculations are $\delta_2 = 0.4a_d^*$, $\delta_3 = 0.4a_d^*$, $\delta_1 = 0.6a_d^*$, and $\delta_3 = 0.4a_d^*$ for Fig. 13a, and $\delta_1 = 0.6a_d^*$, $\delta_2 = 0.4a_d^*$ for Fig. 13b, c.

Figure 13a demonstrates two distinct behaviors of the OS: for core radii less than $\delta_1 \leq 0.6a_d^*$, the OS augments as the shallow donor defect progresses from the structure's heart towards the well region. However, for core radii greater than or equal to $\delta_1 \geq 0.6a_d^*$, the OS reaches its highest-point when the impurity is placed at the core center.

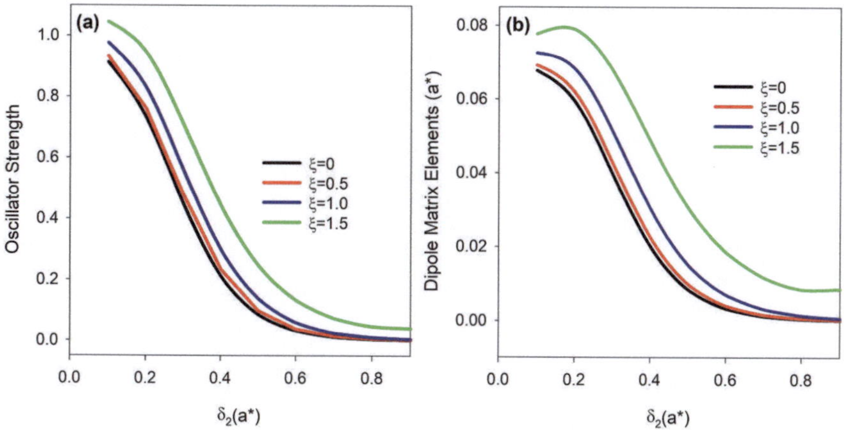

Fig. 12 The OS (**a**) and the dipole matrix element (**b**) versus the δ_2 for several values of the EF and for $\delta_1 = 0.6a_d^*$, $\delta_3 = 0.4a_d^*$ and, $Z = 0$

Furthermore, from Fig. 13b, it can be noted that the OS is significant only for low values of the δ_2. As the δ_2 rises, the OS becomes nearly zero. This behavior is consistent with the findings in Fig. 12a, where both the OS and DME decrease with increasing δ_2.

These results provide insights into the dependence of the OS on structural parameters, EF intensity, and impurity position. The behavior of the OS can be understood based on the wave function overlap and the confinement of electronic states in different regions of the structure.

Furthermore, the OS rises as the shallow donor defect transitions from the structure's heart to the well area. This phenomenon occurs because the overlapping of the wave functions raises when the defect is located near the left border of the well region. Consequently, two wave functions become restricted to the well area, as depicted in Fig. 4g, h. In Fig. 13c, the variations of the OS are displayed for different δ_3 values versus the impurity position. In all cases, it is consistently observed that the OS constantly achieves its highest value when the impurity is positioned at the left border of the well zone. This indicates that the overlapping of the wave functions is greater in this configuration, resulting in higher optical ACs, as evidenced in Fig. 9.

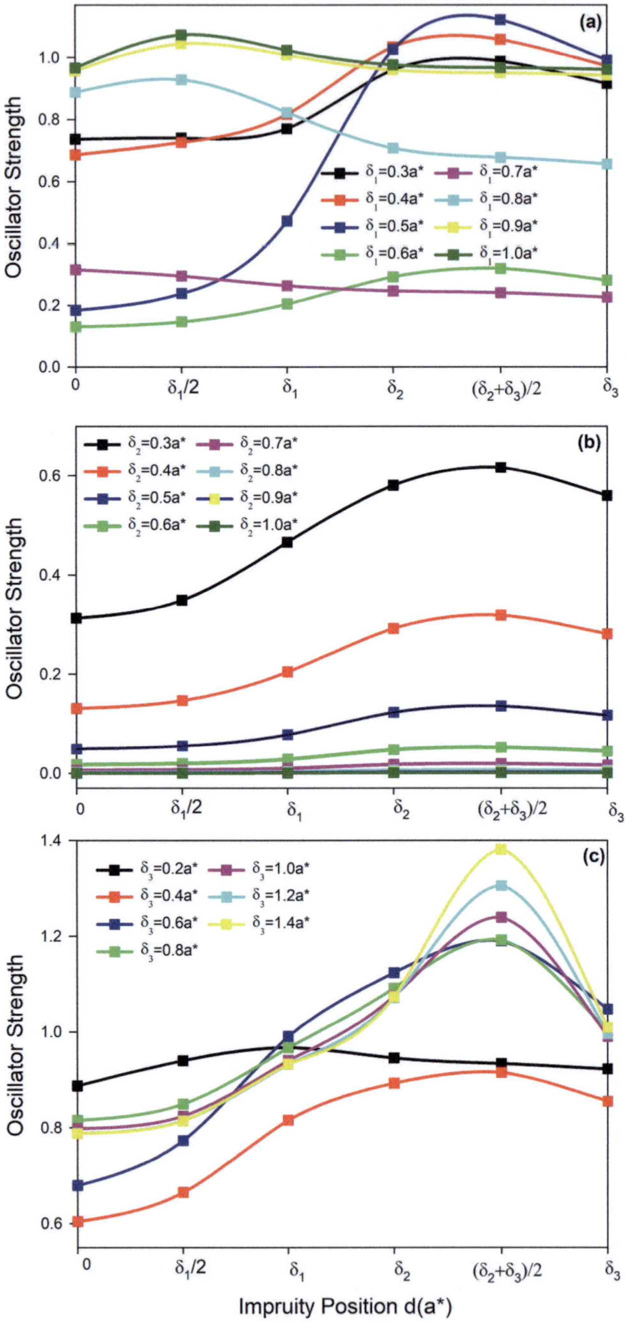

Fig. 13 The OS is plotted against the impurity location for different values of δ_1 (**a**), δ_2 (**b**), and δ_3 (**c**)

4 Conclusion

In this study, We have investigated the linear, third-order nonlinear, and total optical absorption coefficients, in a $GaAs/Ga_{1-x}Al_xAs$ MSQD. In the framework of the EMA, by modelling the confinement potential by a Kratzer potential form, the ground and excited states eigenvalues and their wave functions are determined theoretically by utilizing the FEM to solve the system's Schrödinger equation. Furthermore, in the last section, we examined the impact of EF and impurity position on the ACs and OS.

Our key findings are summarized as follows:

(i) The existence of an impurity in the center causes a shift toward shorter wavelengths in the resonance peaks of ACs, as well as a drop in their intensity.
(ii) The peak position of ACs shifts as the donor defect is relocated from the center to the left border of the well area, the peak position of ACs experiences a shift towards longer wavelengths (red shift), along with an increase in magnitude.
(iii) In the scenario of a large δ_1, the existence and position of the impurity have minimal impact on the maximum point of $\alpha_T(\omega, I)$.
(iv) For tiny barrier width δ_2, the presence of the impurity leads to an enhanced in the $\alpha_T(\omega, I)$ maximum value.
(v) The resonant peak magnitude can either increase or decrease with the EF, depending on the values of the δ_1.
(vi) The peak position can exhibit either a blue or red shift with the EF, depending on the values of the δ_1.
(vii) While the OS augments with the EF for both small and large δ_1 values, the EF's effect is not significant at certain points.
(viii) The OS augments as the donor defect position varies with the shell thickness, but the behavior of the OS becomes intricate when considering the δ_1 and δ_3.

Furthermore, we anticipate that these results will inspire further experimental investigations in this area, providing valuable insights for device applications based on multilayer quantum dots.

References

1. D. Saikia, P. Dutta, N.S. Sarma, N.C. Adhikary, CdTe/ZnS core/shell quantum dot-based ultrasensitive PET sensor for selective detection of Hg (II) in aqueous media. Sens. Actuators B Chem. **230**, 149–156 (2016)
2. P.V. Kamat, Quantum dot solar cells. The next big thing in photovoltaics. J. Phys. Chem. Lett. **4**, 908–918 (2013)
3. S. Adhikary, S. Chakrabarti, Spectral broadening due to post-growth annealing of a long-wave InGaAs/GaAs quantum dot infrared photodetector with a quaternary barrier layer. Thin Solid Films **552**, 146–149 (2014)
4. R. Cheriton, *Design and Characterization of InGaN/GaN Dot-in-Nanowire Heterostructures for High Efficiency Solar Cells* (University of Ottawa, 2018)

5. A. Fakkahi, M. Kirak, A. Sali, Effect of impurity position and electric field on the optical absorption coefficients and oscillator strength in spherical multilayer quantum dot. Eur. Phys. J. Plus **137**, 1068 (2022)
6. A. Fakkahi, H. Dakhlaoui, A. Sali, M. Jaouane, R. Arraoui, K. El-Bakkari, A. Ed-Dahmouny, Pressure, temperature and electric field effects on the photoionization cross section in a multilayered spherical quantum dot. Eur. Phys. J. Plus **137**, 1244 (2022)
7. S. Aktas, F.K. Boz, The binding energy of hydrogenic impurity in multilayered spherical quantum dot. Phys. E **40**, 753–758 (2008)
8. F.K. Boz, S. Aktas, A. Bilekkaya, S.E. Okan, Geometric effects on energy states of a hydrogenic impurity in multilayered spherical quantum dot. Appl. Surf. Sci. **255**, 6561–6564 (2009)
9. F. Boz, S. Aktas, A. Bilekkaya, S. Okan, The multilayered spherical quantum dot under a magnetic field. Appl. Surf. Sci. **256**, 3832–3836 (2010)
10. C.-Y. Hsieh, D.-S. Chuu, C.-Y. Hsieh, Donor states in a multi-layered quantum dot. J. Phys. Condens. Matter **12**, 8641–8653 (2000)
11. A.R. Jafari, Y. Naimi, Linear and nonlinear optical properties of multi-layered spherical nano-systems with donor impurity in the center. J. Comput. Electron. **12**, 36–42 (2013)
12. M. Jaouane, A. Sali, A. Fakkahi, R. Arraoui, A. Ed-Dahmouny, F. Ungan, Photoionization cross section of donor single dopant in multilayer quantum dots under pressure and temperature effects. Phys. E Low-Dimens. Syst. Nanostruct. **144**, 115450 (2022)
13. A. Fakkahi, A. Sali, M. Jaouane, R. Arraoui, A. Ed-Dahmouny, Investigation of the nonlinear optical rectification coefficient in a multilayered spherical quantum dot. Opt. Mater. **132**, 112752 (2022)
14. S. Akgül, M. Şahin, K. Köksal, A detailed investigation of the electronic properties of a multi-layer spherical quantum dot with a parabolic confinement. J. Lumin. **132**, 1705–1713 (2012)
15. A. Emre Kavruk, M. Sahin, F. Koc, Linear and nonlinear optical properties of GaAs/Al_xGa_{1-x}As/GaAs/Al_yGa_{1-y}As multi-shell spherical quantum dot. J. Appl. Phys. **114**, 183704 (2013)
16. M. Jaouane, A. Sali, A. Ezzarfi, A. Fakkahi, R. Arraoui, Study of hydrostatic pressure, electric and magnetic fields effects on the donor binding energy in multilayer cylindrical quantum dots. Phys. E **127**, 114543 (2021)
17. K. Batra, V. Prasad, Spherical quantum dot in Kratzer confining potential: study of linear and nonlinear optical absorption coefficients and refractive index changes. Eur. Phys. J. B **91**, 298 (2018)
18. F. Ungan, J. Martínez-Orozco, R. Restrepo, M. Mora-Ramos, The nonlinear optical properties of GaAs-based quantum wells with Kratzer–Fues confining potential: role of applied static fields and non-resonant laser radiation. Optik **185**, 881–887 (2019)
19. A. Dehyar, G. Rezaei, A. Zamani, Electronic structure of a spherical quantum dot: effects of the Kratzer potential, hydrogenic impurity, external electric and magnetic fields. Phys. E **84**, 175–181 (2016)
20. A. Fakkahi, A. Sali, M. Jaouane, R. Arraoui, A. Ed-Dahmouny, Study of photoionization cross section and binding energy of shallow donor impurity in multilayered spherical quantum dot. Phys. E Low-Dimens. Syst. Nanostruct. 115351 (2022)
21. D. Baghdasaryan, E. Kazaryan, H. Sarkisyan, Photoionization and electrostatic multipoles properties of spherical core/shell/shell quantum nanolayer with off-center impurity. Superlattices Microstruct. **104**, 69–77 (2017)
22. A. Fakkahi, M. Jaouane, K. Limame, A. Sali, M. Kirak, R. Arraoui, A. Ed-Dahmouny, K. El-Bakkari, H. Azmi, The impact of the electric field on the photoionization cross section, polarizability, and donor impurity binding energy in multilayered spherical quantum dot. Appl. Phys. A **129**, 188 (2023)
23. S. M'zerd, M. El Haouari, M. Aghoutane, M. El-Yadri, E. Feddi, F. Dujardin, I. Zorkani, A. Jorio, M. Sadoqi, G. Long, Electric field effect on the photoionization cross section of a single dopant in a strained AlAs/GaAs spherical core/shell quantum dot. J. Appl. Phys. **124**, 164303 (2018)

24. R. Arraoui, E.-D. Ayoub, A. Sali, K. El-Bakkari, M. Jaouane, A. Fakkahi, Tridimensional electric field effect on diamagnetic susceptibility and polarisability of a donor impurity in a double quantum dot. Philos. Mag. 1–16 (2023)
25. M. Jaouane, A. Fakkahi, A. Ed-Dahmouny, K. El-Bakkari, A.T. Tuzemen, R. Arraoui, A. Sali, F. Ungan, Modeling and simulation of the influence of quantum dots density on solar cell properties. Eur. Phys. J. Plus **138**, 148 (2023)
26. A. Ed-Dahmouny, A. Sali, N. Es-Sbai, R. Arraoui, M. Jaouane, A. Fakkahi, K. El-Bakkari, C.A. Duque, Combined effects of hydrostatic pressure and electric field on the donor binding energy, polarizability, and photoionization cross-section in double GaAs/$Al_xGa_{1-x}As$ quantum dots. Eur. Phys. J. B **95**, 136 (2022)
27. A. Ed-Dahmouny, A. Sali, N. Es-Sbai, R. Arraoui, C.A. Duque, The impact of hydrostatic pressure and temperature on the binding energy, linear, third-order nonlinear, and total optical absorption coefficients and refractive index changes of a hydrogenic donor impurity confined in GaAs/$Al_xGa_{1-x}As$ double quantum dots. Eur. Phys. J. Plus **137**, 784 (2022)
28. A. Sali, J. Kharbach, A. Rezzouk, M. Ouazzani Jamil, The effects of polaronic mass and conduction band non-parabolicity on a donor binding energy under the simultaneous effect of pressure and temperature basing on the numerical FEM in a spherical quantum dot. Superlattices Microstruct. **104**, 93–103 (2017)
29. M. Şahin, F. Tek, A. Erdinç, The photoionization cross section of a hydrogenic impurity in a multi-layered spherical quantum dot. J. Appl. Phys. **111**, 084317 (2012)
30. A. Sali, A. Rezzouk, N. Es-Sbai, M.O. Jamil, The simultaneous effects of the wetting layer, intense laser and the conduction band non-parabolicity on the donor binding energy in a InAs/GaAs conical quantum dot using the numerical FEM. Indian J. Pure Appl. Phys. **57**, 9 (2019)
31. A. Sali, H. Satori, M. Fliyou, H. Loumrhari, The photoionization cross-section of impurities in quantum dots. Phys. Stat. Sol. (b) **232**, 209–219 (2002)
32. C. Duque, E. Kasapoglu, S. Şakiroglu, H. Sari, I. Sökmen, Intense laser effects on nonlinear optical absorption and optical rectification in single quantum wells under applied electric and magnetic field. Appl. Surf. Sci. **257**, 2313–2319 (2011)
33. G. Rezaei, M. Vahdani, B. Vaseghi, Nonlinear optical properties of a hydrogenic impurity in an ellipsoidal finite potential quantum dot. Curr. Appl. Phys. **11**, 176–181 (2011)
34. H. Dakhlaoui, Tunability of the optical absorption and refractive index changes in step-like and parabolic quantum wells under external electric field. Optik **168**, 416–423 (2018)
35. L.R. Ram-Mohan, S. Saigal, D. Dossa, J. Shertzer, The finite-element method for energy eigenvalues of quantum mechanical systems. Comput. Phys. **4**(1), 50 (1990)
36. H. Satori, A. Sali, The finite element simulation for the shallow impurity in quantum dots. Phys. E **48**, 171–175 (2013)
37. O.C. Zienkiewicz, R.L. Taylor, J.Z. Zhu, *The Finite Element Method: Its Basis and Fundamentals*, 6th edn. (Elsevier, Amsterdam Heidelberg, 2010)
38. A. Sali, H. Satori, The combined effect of pressure and temperature on the impurity binding energy in a cubic quantum dot using the FEM simulation. Superlattices Microstruct. **69**, 38–52 (2014)
39. A. Özmen, Y. Yakar, B. Çakır, Ü. Atav, Computation of the oscillator strength and absorption coefficients for the intersubband transitions of the spherical quantum dot. Opt. Commun. **282**, 3999–4004 (2009)
40. H. Taş, M. Şahin, The inter-sublevel optical properties of a spherical quantum dot-quantum well with and without a donor impurity. J. Appl. Phys. **112**, 053717 (2012)

Kratzer Potential's Impact on Optical Absorption in Multilayer Quantum Dots

A. Fakkahi, H. Azmi, M. Jaouane, A. Sali, A. Ed-Dahmouny, K. El-Bakkari, R. Arraoui, and S. Elotmani

1 Introduction

The utilization of low-dimensional structures in optical devices, sensors [1], and optoelectronics has garnered significant attention in recent years. Examples of such structures include quantum dots (QDs), quantum wells (QWs), quantum wire wells (QWWs), and multilayer quantum dots (MQDs). Advances in technology have enabled the fabrication of QDs and MQDs [1–3]. Researchers have shown great interest in determining the optical properties of these nanostructures, as this knowledge is crucial for designing efficient optoelectronic devices [1–4].

Multilayer structures are known to confine electrons within multiple regions, where the electron experiences varying confinement potentials in each layer [5–8]. The application of MQDs in optoelectronic devices such as LEDs and optical amplifiers has increased recently, due to their superior experimental performance. Theoretical studies predicting the optical behavior of MQDs are vital, as these structures play a key role in developing novel, high-efficiency optical devices [7, 9–11].

It is important to note that external factors—including temperature, pressure, and impurities—can significantly influence the optical properties of QDs and MQDs. Consequently, this field has attracted extensive research. For example, Jaouane et al. [12] investigated how pressure and temperature affect the photoionization cross-section (PCS) in cylindrical MQDs. Similarly, Fakkahi et al. [13] examined optical rectification in a multilayer spherical quantum dot (MSQD) using the finite

A. Fakkahi (✉) · H. Azmi · M. Jaouane · A. Sali · K. El-Bakkari · R. Arraoui · S. Elotmani
Laboratory of Solid-State Physics, Department of Physics, Faculty of Science Dhar El Mahraz, Sidi Mohamed Ben Abdellah University, Fez, Morocco
e-mail: abdelghani.fakkahi@usmba.ac.ma

A. Ed-Dahmouny
Science and Engineering Research Laboratory (LaRSI), Faculty of Sciences and Technology, Sidi Mohamed Ben Abdellah University, Fez, Morocco

R. En-nadir (ed.), *Quantum Systems at the Nanoscale*, Lecture Notes in Nanoscale Science and Technology 45, https://doi.org/10.1007/978-3-032-06645-9_11

element method (FEM). Readers interested in further details about MQDs may consult [7–10, 12, 14–16].

To theoretically analyze the physical properties of QDs and MQDs, knowledge of the confining potential is essential. Various potentials have been employed in literature, including Kratzer, Morse, Deng–Fan–Eckart, and Woods–Saxon (WS) potentials [17–19]. Numerous researchers have studied the physical characteristics of QDs and MQDs under these different potentials [10, 12, 14, 15, 20–34]. When the Schrödinger equation cannot be solved analytically for certain potentials, approximation methods such as perturbation theory, diagonalization, factorization, and semi-linearization are applied.

One effective numerical approach for solving the Schrödinger equation (SE) is the Finite Element Method (FEM), which finds extensive use in fields like physics, optics, electrochemistry, and plasma studies. In FEM, the space is discretized into numerous small meshes, each treated as an individual element. The governing differential equations are then applied to these elements, allowing the variables to be computed throughout the domain [35–38]. Using FEM, the SE is solved for a multilayer spherical quantum dot within the effective mass approximation (EMA). This process yields eigenvalues and eigenstates, from which absorption coefficients (ACs) can be calculated.

2 Theoretical Framework and Numerical Methods

2.1 *Electronic Properties*

We examine the confinement of an electron within a MSQD composed of $GaAs/AlGaAs/GaAs/AlGaAs$ layers, with a radius denoted as R_1. In the presence of a uniform static EF along the z-direction, we utilize the EMA to express the Hamiltonian of the considered system can be written in spherical coordinate systems as stated below:

$$\Big[\frac{1}{m^*}\left(-\frac{\partial^2}{\partial r^2}+\frac{2}{r}\frac{\partial}{\partial r}+\frac{1}{r^2}\frac{\partial^2}{\partial \theta^2}\frac{\cot\theta}{r^2}\frac{\partial}{\partial\theta}+\frac{m^2}{r^2\sin^2\theta}\right) + V(r)-\frac{2Z}{\varepsilon\sqrt{r^2+d^2-2rd\cos\theta}}\Big]\psi(r,\theta)=E\psi(r,\theta), \tag{1}$$

Here, m^* represents the effective mass of the electron, ε denotes the dielectric constant, Z represents the charge of the impurity, and r and d correspond to the position coordinates of the electron and impurity, respectively.

The effective masses (dielectric constants) of materials are given by

$$m^*(\varepsilon)=\begin{cases} m_1^*(\varepsilon_1) & \text{for} \quad 0\le r\le R_1 \quad \text{and} \quad R_2\le r\le R_3 \\ m_2^*(\varepsilon_2) & \text{for} \quad R_1<r<R_2 \quad \text{and} \quad r>R_3 \end{cases} \tag{2}$$

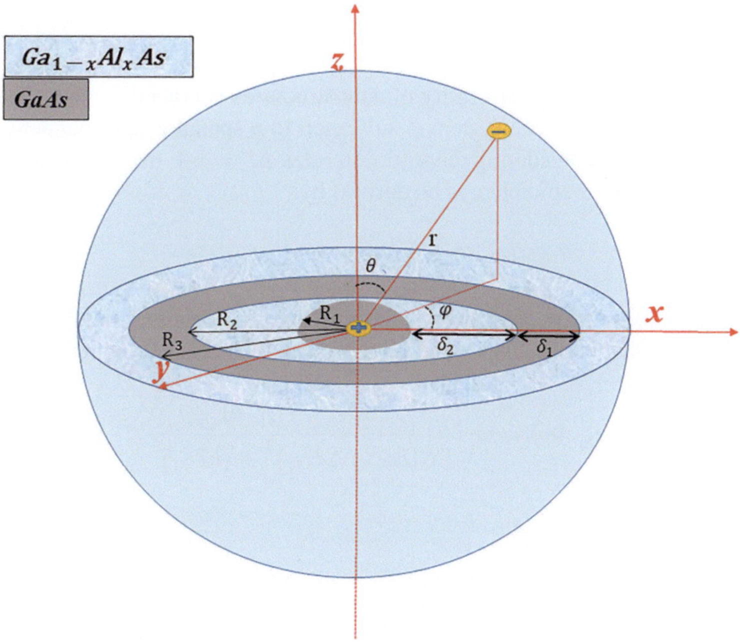

Fig. 1 A schematic view of the MSQD

where R_1 is the radius of core, R_2 and R_3 are inner and outer shell regions, respectively. $m_1^*(\varepsilon_1)$ and $m_2^*(\varepsilon_2)$ are the electron effective masses (dielectric constants) in $GaAs$ dot and $AlGaAs$ barrier materials, respectively.

The confining potential, denoted as $V(r)$, is defined as the sum of two components: the Kratzer–Fues potential (V_K) and the rectangular potential (V_i). The mathematical expression of the V_K is as follows:

$$V_K = -D\left(\frac{2R_0}{r} - \frac{R_0^2}{r^2}\right) \tag{3}$$

The rectangular potential V_i,

$$V_i = \begin{cases} 0 & \text{for} \quad 0 \le r \le R_1 \quad \text{and} \quad R_2 \le r \le R_3 \\ V_0 & \text{for} \quad R_1 < r < R_2 \quad \text{and} \quad r > R_3 \end{cases} \tag{4}$$

The graphical representations of Kratzer–Fues potential profiles are presented for different parameter values, and $V_0 = 228,$ meV is the confinement potential height, which is serves as a characteristic value for D. Figure 1 reveals the form of schematic representation of the studied structure.

2.2 Optical Properties

Investigating the optical nonlinearity of nanostructures is crucial for their potential applications in devices. The system is subjected to a spatial electromagnetic wave of frequency ω, and considering the selection rules $\Delta l = \pm 1$, the analytical form of the total absorption coefficient can be derived by:

$$\alpha_T(\omega, I) = \alpha_L^{(1)}(\omega) + \alpha_N^{(3)}(\omega, I) \tag{5}$$

The above expression comprises two terms, the first denoting the linear optical absorption coefficient and the second representing the nonlinear optical absorption coefficient.

$$\alpha_L^{(1)}(\omega) = \omega\sqrt{\frac{\mu}{\varepsilon_R}}\frac{\sigma_V e^2 \left|M_{fi}\right|^2 \hbar\Gamma_{fi}}{\left(\hbar\omega - \Delta E_{fi}\right)^2 + \left(\hbar\Gamma_{fi}\right)^2}, \tag{6}$$

and

$$\alpha_N^{(3)}(\omega, I) = \omega\sqrt{\frac{\mu}{\varepsilon_R}}\left(\frac{I}{2\varepsilon_0 n_r c}\right)\frac{\sigma_V \left|M_{fi}\right|^2 \hbar\Gamma_{fi}}{\left[\left(\hbar\omega - \Delta E_{fi}\right)^2 + \left(\hbar\Gamma_{fi}\right)^2\right]^2}\left\{4\left|M_{fi}\right|^2 - \frac{\left(M_{ff} - M_{ii}\right)^2\left[3E_{fi}^2 - 4\Delta E_{fi}\hbar\omega + \hbar^2(\omega^2 - \Gamma_{if}^2)\right]}{E_{fi}^2 + (\hbar\Gamma_{if})^2}\right\}. \tag{7}$$

In the given context, the symbols used have the following meanings: σ_V denotes the electron density, μ represents the magnetic permeability, ε_0 corresponds to the dielectric permittivity of vacuum, ε_R stands for the real part of the relative permittivity, ω represents the angular frequency of the incident photon, $\hbar\omega$ signifies the energy of the incident photon, n_r denotes the refractive index of the dot material, c represents the speed of light, e corresponds to the electron charge, I signifies the intensity of the incident light, E_i and E_f represent the initial and final state energies, respectively, ΔE_{fi} represents the energy difference between the final and initial states, Γ_{if} is the relaxation rate for states i and f, τ denotes the relaxation time, and M_{if} stands for the dipole transition matrix element between states i and f. In this study, we have considered polarized electromagnetic radiation in the z-direction.

3 Results and Interpretation

In this section, we provide the numerical findings of the effects of Kratzer–Fues potential parameters on the ACs as well as on the transition energy and the square dipole elements matrix for different allowed transitions. The numerical parameters used in our simulations are taken as: $x = 0.3$ ($V_0 = 228$ meV), $a_d^* = 103.54$ Å and $R_y^* = 5.305$ meV.

3.1 The Impact of the Parameter R_0 on the Optical ACs

Figure 2 presents the changes of the energy differences (a) and the square dipole matrix elements $M_{f,i}^2$ (b) for $1p - 2s$ and $2s - 2p$ transitions versus the QD radius. Transition energy ($E_{f,i}$) is an important factor because it can modify the optical characteristics of quantum nanostructures. As expected, an increment in parameter a leads to a decrease in all energy levels. This is due to weakening of the quantum geometric confinement. However, these reduction rates are different for each energy level. The reduction rate for the $1p$ level is approximately linear. However, the energy values decrease rapidly up to $R_0 = 0.8a_d^*$ for the 2s level and $R_0 = 0.6a_d^*$ for the $2p$ level. Therefore, we can say that transition energy for $1p - 2s$ transition decreases monotonously. But, the $2s - 2p$ transition energy increases until $R_0 = 0.4a_d^*$ and then decreases for larger values of R_0, so its behavior is not monotonous. One of the important factors determining the behavior of the optical characteristics is the ($M_{f,i}^2$), which defines the transition between the initial and final states. The dependencies of the square dipole matrix elements on the parameter R_0 for $1p - 2s$ and $2s - 2p$ transitions are displayed in Fig. 2b. The $M_{f,i}^2$ is increasing (decreasing) almost linearly with increasing the parameter R_0 for $1p - 2s(2s - 2p)$ transition. We can explain the physical reason of these behaviors by the fact that, for extremely small parameter values of R_0, the wave functions of $1p$ ($2p$) and 2s states are confined in the separate (same) zones and so, the overlapping of the wave functions, i.e., the $M_{f,i}^2$ becomes smaller (larger). However, the $M_{f,i}^2$ augments (diminishes) with further increasing of the parameter a value due to the fact that the wave functions are localized in the same (different) regions.

Figure 3 illustrates the influence of the $\hbar\omega$ on the $\alpha_L^{(1)}(\omega)$, $\alpha_N^{(3)}(\omega, I)$ and $\alpha_T(\omega, I)$ for $2s - 2p$ transition for three values of the parameter R_0 (0.4, 0.6 and $0.8a_d^*$). In this figure, the dotted, dashed, solid lines represent the linear, nonlinear and total ACs curves, respectively. It is evident from the figure that the ACs peaks have shifted toward lower photon energy values as the parameter a increases. The transition energy determines the peak position and the physical reason of this shifting of the absorption peaks is that the transition energy reduces with the enlargement of the parameter R_0. In addition, the strength of the ACs decreases with the rise in the value of parameter R_0. These results are consistent with the variations of transition energy and $M_{f,i}^2$ plotted in Fig. 2a, b, respectively.

To comprehend the impact of the parameter R_0 on the linear and nonlinear optical characteristics for $1p - 2s$ transition, we have investigated in Fig. 4 the changes of the ACs against $\hbar\omega$ for various values of the factor R_0. The curve of the ACs for $1p - 2s$ transition plotted in Fig. 4 exhibits red-shift of the peak position as a becomes higher in a similar manner to $2s - 2p$ transition. It is evident from the figure that there is an increment in the amplitude of ACs which comes from of the

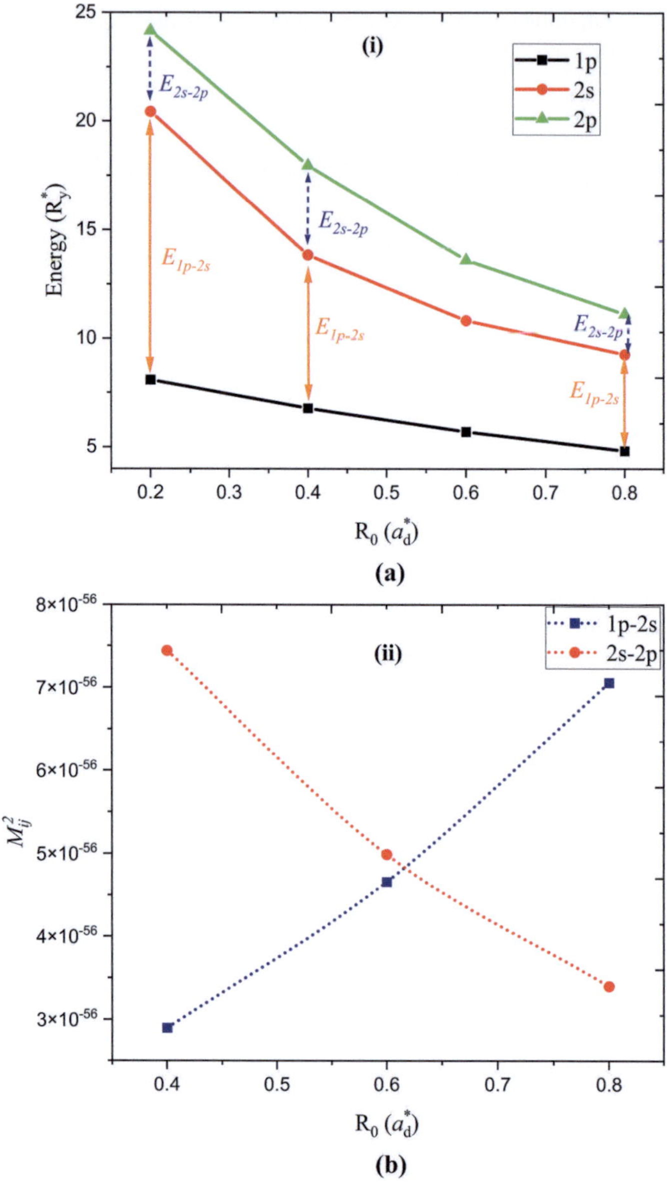

Fig. 2 Transition energy (**a**) and the square element matrix (**b**) versus the parameter a for different transition $2s - 2p$, and $1p - 2s$

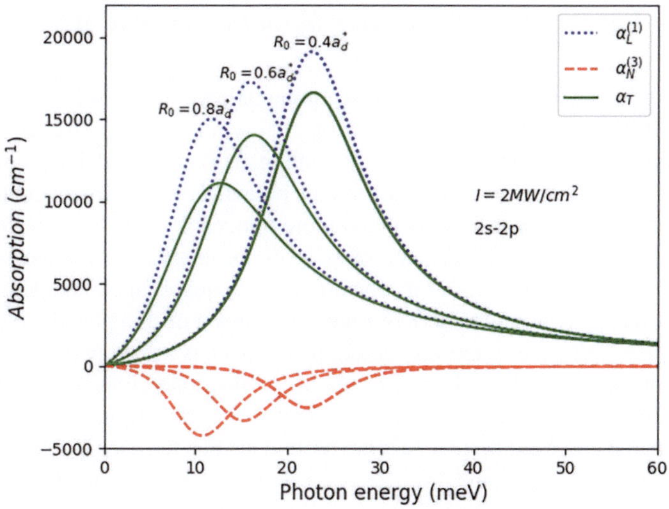

Fig. 3 The $\alpha_L^{(1)}(\omega)$, $\alpha_N^{(3)}(\omega, I)$ and $\alpha_T(\omega, I)$ versus $\hbar\omega$ for different values of R_0 for $2s - 2p$ transition

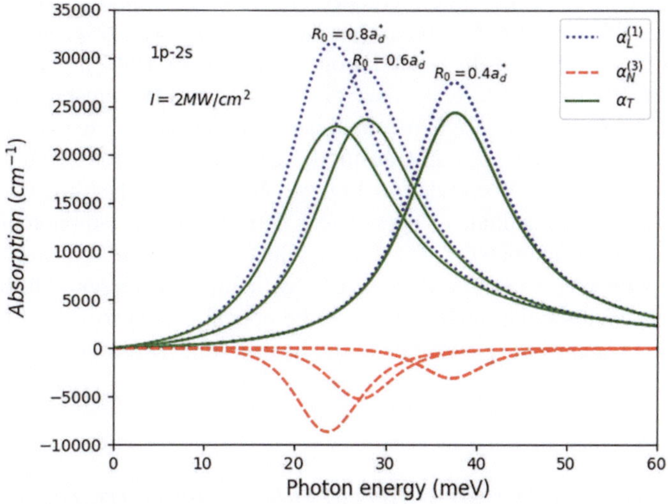

Fig. 4 The $\alpha_L^{(1)}(\omega)$, $\alpha_N^{(3)}(\omega, I)$ and $\alpha_T(\omega, I)$ against the $\hbar\omega$ for different values of the factor R_0 for $1p - 2s$ transition

increase in the parameter R_0. This is due to the increase in the $M_{f,i}^2$ for this transition, as explained in Fig. 2b.

3.2 The Impact of the Parameter N on the Optical ACs

To visualize the impact of the depth of the potential, we have plotted, respectively, in Fig. 5a, b the behaviors of the transition energy and the $M_{f,i}^2$ for $2s - 2p$ and $1p - 2s$ transitions versus the factor N. When Fig. 5a is examined, it is found that all energy levels ($1p$, $2s$ and $2p$) increase with the potential depth. Moreover, the energy difference for $1p - 2s$ transition is much more noticeable than $2s - 2p$ because of the diminution in energy difference between levels as thc quantum number increases. It is possible to notice that for both transitions, the square dipole element matrix's is getting smaller as N augments. The physical reason of this phenomenon is that the overlapping of wave functions for small N values is more significant than that obtained for large N values. But, it needs to be mentioned that the square dipole matrix element increases in case of $N > 3$ for $1p - 2s$ transition.

In Fig. 6, the curves of optical ACs are plotted against $\hbar\omega$ energy for three values of the parameter $N = 2$, 3 and 4 for $2s - 2p$ with $I = 5\,\mathrm{MW/cm^2}$. It is clear from this graph that the parameter N induces a displacement of the ACs peak towards lower energies (red shift). The reason for this shift can be understood by considering the dependence of the $2s - 2p$ transition energy on the potential depth as depicted in Fig. 5a. Moreover, it is found that a decrement in the amplitude of ACs results from the dependence of the $M_{f,i}^2$ on the parameter N as shown in Fig. 5b. To evaluate the effects of the depth of the potential on the optical ACs of $1p - 2s$ transition, we have plotted Fig. 7 for several values of the factor N. In contrast to the previous transition (Fig. 6), the increase in potential depth causes the ACs to undergo a blue-shift of the ACs for the $1p - 2s$ transition, due to the variation of the transition energy with the factor N revealed in Fig. 5a. On the other hand, it is important to observe that the magnitude of optical ACs does not follow a monotonic pattern with respect to the parameter N. We find that the peak values of ACs decrease with increasing parameter N and then they increase with further increases of the parameter N. To achieve substantial modifications in the optical characteristics of MSQD, it is advisable to employ the parameters of the Kratzer–Fues potential, namely R_0 and N.

3.3 The Optical ACs for Different Intersubband Transitions

The curves of the optical ACs against $\hbar\omega$ for $1s - 1p$, $1s - 2p$, and $1s - 3p$ transitions are plotted in Fig. 8. The same figure is plotted in Fig. 9 for $1d - 1f$, $1d - 2f$, and $1d - 3f$ transitions by keeping all parameters constant. From Fig. 8 one can see that the ACs peak exhibits blue-shift as the excited states quantum number increases from $1p$ to $3p$. This phenomenon occurs due to the larger energy difference between the ground state and a smaller principal quantum number n in comparison to the energy difference between the ground state and a higher principal quantum number n. Moreover, the peak amplitude of the ACs increases as n of p-orbital (l = 1)

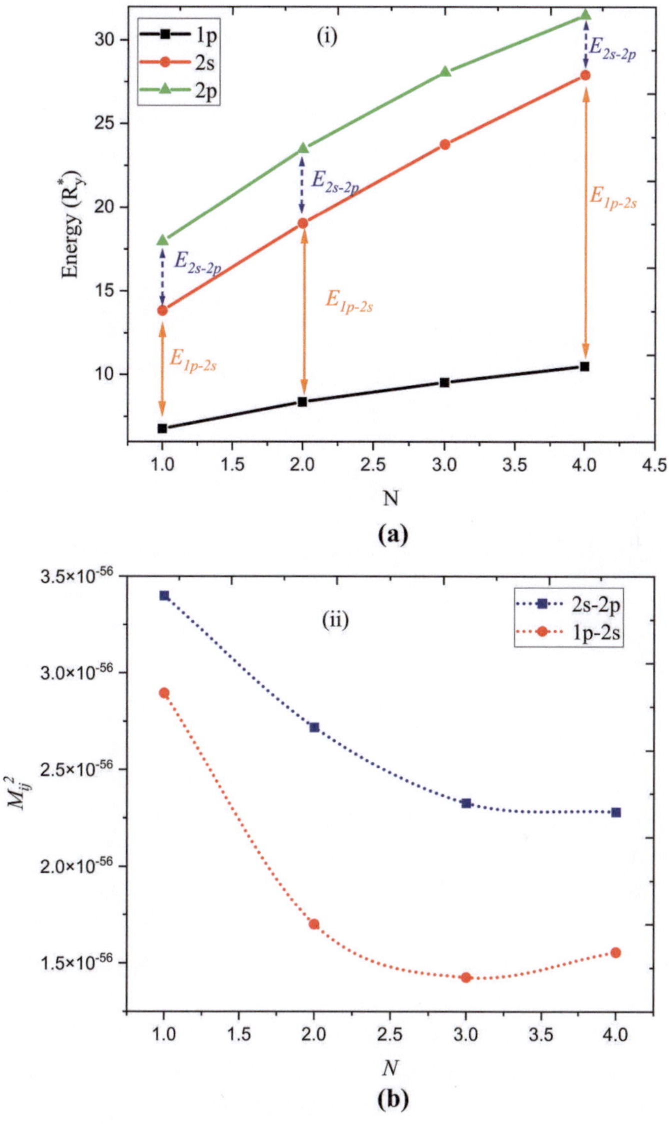

Fig. 5 Transition energy (**a**) and the square element matrix (**b**) versus the parameter N for different transition $2s - 2p$, and $1p - 2s$

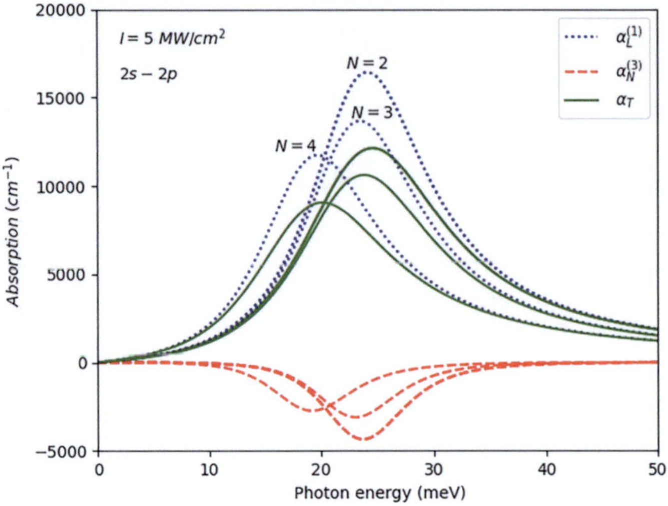

Fig. 6 The $\alpha_L^{(1)}(\omega)$, $\alpha_N^{(3)}(\omega, I)$ and $\alpha_T(\omega, I)$ against the $\hbar\omega$ for different values of the variable N for $2s - 2p$ transition

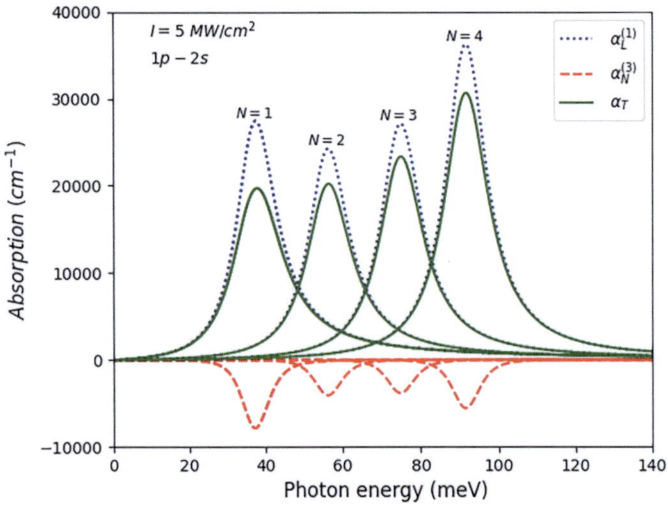

Fig. 7 The $\alpha_L^{(1)}(\omega)$, $\alpha_N^{(3)}(\omega, I)$ and $\alpha_T(\omega, I)$ against the $\hbar\omega$ for different values of the variable N for $1p - 2s$ transition

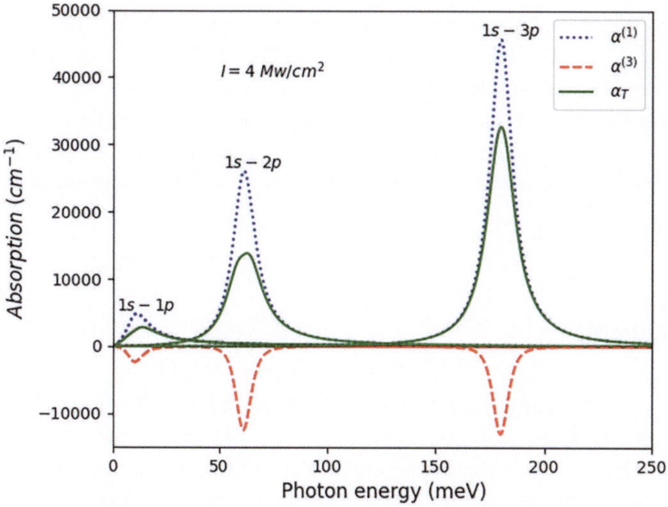

Fig. 8 The $\alpha_L^{(1)}(\omega)$, $\alpha_N^{(3)}(\omega, I)$ and $\alpha_T(\omega, I)$ versus the $\hbar\omega$ for different transition $1s-1p$, $1s-2p$, and $1s-3p$

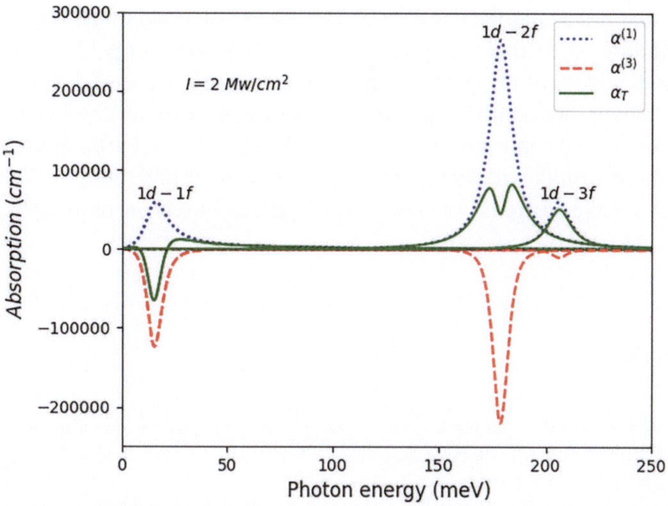

Fig. 9 The $\alpha_L^{(1)}(\omega)$, $\alpha_N^{(3)}(\omega, I)$ and $\alpha_T(\omega, I)$ against the $\hbar\omega$ for different transition $1d-1f$, $1d-2f$, and $1d-3f$

increments. We should also mention that the behavior of the ACs peak position for $1d - 1f$, $1d - 2f$, and $1d - 3f$ transitions in Fig. 9 follow the same general trend as it plotted in Fig. 8. In addition, the amplitude of the resonance peak augments when the quantum number increases from $1f$ to $2f$and then it followed by a decreases for the transition $1d - 3f$. The aforementioned outcome can be attributed to the increased overlap of the wave functions of $1d$ and $2f$ that are confined in the same zone.

4 Conclusion

In this study, We have investigated the linear, third-order nonlinear, and total optical absorption coefficients, in a $GaAs/Ga_{1-x}Al_xAs$ MSQD. In the framework of the EMA, by modelling the confinement potential by a Kratzer potential form, the ground and excited states eigenvalues and their wave functions are determined theoretically by utilizing the FEM to solve the system's Schrödinger equation. Moreover, in this paper, we examined the ACs against the excitation photon energy for two allowed transitions, $1p - 2s$ and $2s - 2p$. The findings indicate that the Kratzer confinement potential has a significant effect on the optical ACs and transition energy. Our essential findings show that the optical ACs and transition energy are sensitive to the changes in the depth of the potential Kratzer quantum confinement and to the core radius. Moreover, our results show that the optical ACs versus the $\hbar\omega$ are strongly dependent on the subband state where the electron is excited. Furthermore, we anticipate that these results will inspire further experimental investigations in this area, providing valuable insights for device applications based on multilayer quantum dots.

References

1. D. Saikia, P. Dutta, N.S. Sarma, N.C. Adhikary, CdTe/ZnS core/shell quantum dot-based ultrasensitive PET sensor for selective detection of Hg (II) in aqueous media. Sens. Actuators B Chem. **230**, 149–156 (2016)
2. P.V. Kamat, Quantum dot solar cells. The next big thing in photovoltaics. J. Phys. Chem. Lett. **4**, 908–918 (2013)
3. S. Adhikary, S. Chakrabarti, Spectral broadening due to post-growth annealing of a long-wave InGaAs/GaAs quantum dot infrared photodetector with a quaternary barrier layer. Thin Solid Films **552**, 146–149 (2014)
4. R. Cheriton, *Design and Characterization of InGaN/GaN Dot-in-Nanowire Heterostructures for High Efficiency Solar Cells* (University of Ottawa, 2018)
5. A. Fakkahi, M. Kirak, A. Sali, Effect of impurity position and electric field on the optical absorption coefficients and oscillator strength in spherical multilayer quantum dot. Eur. Phys. J. Plus **137**, 1068 (2022)
6. A. Fakkahi, H. Dakhlaoui, A. Sali, M. Jaouane, R. Arraoui, K. El-Bakkari, A. Ed-Dahmouny, Pressure, temperature and electric field effects on the photoionization cross section in a multilayered spherical quantum dot. Eur. Phys. J. Plus **137**, 1244 (2022)

7. S. Aktas, F.K. Boz, The binding energy of hydrogenic impurity in multilayered spherical quantum dot. Phys. E **40**, 753–758 (2008)
8. F.K. Boz, S. Aktas, A. Bilekkaya, S.E. Okan, Geometric effects on energy states of a hydrogenic impurity in multilayered spherical quantum dot. Appl. Surf. Sci. **255**, 6561–6564 (2009)
9. F. Boz, S. Aktas, A. Bilekkaya, S. Okan, The multilayered spherical quantum dot under a magnetic field. Appl. Surf. Sci. **256**, 3832–3836 (2010)
10. C.-Y. Hsieh, D.-S. Chuu, C.-Y. Hsieh, Donor states in a multi-layered quantum dot. J. Phys. Condens. Matter **12**, 8641–8653 (2000)
11. A.R. Jafari, Y. Naimi, Linear and nonlinear optical properties of multi-layered spherical nano-systems with donor impurity in the center. J. Comput. Electron. **12**, 36–42 (2013)
12. M. Jaouane, A. Sali, A. Fakkahi, R. Arraoui, A. Ed-Dahmouny, F. Ungan, Photoionization cross section of donor single dopant in multilayer quantum dots under pressure and temperature effects. Phys. E Low-Dimens. Syst. Nanostruct. **144**, 115450 (2022)
13. A. Fakkahi, A. Sali, M. Jaouane, R. Arraoui, A. Ed-Dahmouny, Investigation of the nonlinear optical rectification coefficient in a multilayered spherical quantum dot. Opt. Mater. **132**, 112752 (2022)
14. S. Akgül, M. Şahin, K. Köksal, A detailed investigation of the electronic properties of a multi-layer spherical quantum dot with a parabolic confinement. J. Lumin. **132**, 1705–1713 (2012)
15. A. Emre Kavruk, M. Sahin, F. Koc, Linear and nonlinear optical properties of GaAs/Al_xGa_{1-x}As/GaAs/Al_yGa_{1-y}As multi-shell spherical quantum dot. J. Appl. Phys. **114**, 183704 (2013)
16. M. Jaouane, A. Sali, A. Ezzarfi, A. Fakkahi, R. Arraoui, Study of hydrostatic pressure, electric and magnetic fields effects on the donor binding energy in multilayer cylindrical quantum dots. Phys. E **127**, 114543 (2021)
17. K. Batra, V. Prasad, Spherical quantum dot in Kratzer confining potential: study of linear and nonlinear optical absorption coefficients and refractive index changes. Eur. Phys. J. B **91**, 298 (2018)
18. F. Ungan, J. Martínez-Orozco, R. Restrepo, M. Mora-Ramos, The nonlinear optical properties of GaAs-based quantum wells with Kratzer–Fues confining potential: role of applied static fields and non-resonant laser radiation. Optik **185**, 881–887 (2019)
19. A. Dehyar, G. Rezaei, A. Zamani, Electronic structure of a spherical quantum dot: effects of the Kratzer potential, hydrogenic impurity, external electric and magnetic fields. Phys. E **84**, 175–181 (2016)
20. A. Fakkahi, A. Sali, M. Jaouane, R. Arraoui, A. Ed-Dahmouny, Study of photoionization cross section and binding energy of shallow donor impurity in multilayered spherical quantum dot. Phys. E Low-Dimens. Syst. Nanostruct. 115351 (2022)
21. D. Baghdasaryan, E. Kazaryan, H. Sarkisyan, Photoionization and electrostatic multipoles properties of spherical core/shell/shell quantum nanolayer with off-center impurity. Superlattices Microstruct. **104**, 69–77 (2017)
22. A. Fakkahi, M. Jaouane, K. Limame, A. Sali, M. Kirak, R. Arraoui, A. Ed-Dahmouny, K. El-Bakkari, H. Azmi, The impact of the electric field on the photoionization cross section, polarizability, and donor impurity binding energy in multilayered spherical quantum dot. Appl. Phys. A **129**, 188 (2023)
23. S. M'zerd, M. El Haouari, M. Aghoutane, M. El-Yadri, E. Feddi, F. Dujardin, I. Zorkani, A. Jorio, M. Sadoqi, G. Long, Electric field effect on the photoionization cross section of a single dopant in a strained AlAs/GaAs spherical core/shell quantum dot. J. Appl. Phys. **124**, 164303 (2018)
24. R. Arraoui, E.-D. Ayoub, A. Sali, K. El-Bakkari, M. Jaouane, A. Fakkahi, Tridimensional electric field effect on diamagnetic susceptibility and polarisability of a donor impurity in a double quantum dot. Philos. Mag. 1–16 (2023)
25. M. Jaouane, A. Fakkahi, A. Ed-Dahmouny, K. El-Bakkari, A.T. Tuzemen, R. Arraoui, A. Sali, F. Ungan, Modeling and simulation of the influence of quantum dots density on solar cell properties. Eur. Phys. J. Plus **138**, 148 (2023)

26. A. Ed-Dahmouny, A. Sali, N. Es-Sbai, R. Arraoui, M. Jaouane, A. Fakkahi, K. El-Bakkari, C.A. Duque, Combined effects of hydrostatic pressure and electric field on the donor binding energy, polarizability, and photoionization cross-section in double GaAs/Al_xGa_{1-x}As quantum dots. Eur. Phys. J. B **95**, 136 (2022)
27. A. Ed-Dahmouny, A. Sali, N. Es-Sbai, R. Arraoui, C.A. Duque, The impact of hydrostatic pressure and temperature on the binding energy, linear, third-order nonlinear, and total optical absorption coefficients and refractive index changes of a hydrogenic donor impurity confined in GaAs/Al_xGa_{1-x}As double quantum dots. Eur. Phys. J. Plus **137**, 784 (2022)
28. A. Sali, J. Kharbach, A. Rezzouk, M. Ouazzani Jamil, The effects of polaronic mass and conduction band non-parabolicity on a donor binding energy under the simultaneous effect of pressure and temperature basing on the numerical FEM in a spherical quantum dot. Superlattices Microstruct. **104**, 93–103 (2017)
29. M. Şahin, F. Tek, A. Erdinç, The photoionization cross section of a hydrogenic impurity in a multi-layered spherical quantum dot. J. Appl. Phys. **111**, 084317 (2012)
30. A. Sali, A. Rezzouk, N. Es-Sbai, M.O. Jamil, The simultaneous effects of the wetting layer, intense laser and the conduction band non-parabolicity on the donor binding energy in a InAs/GaAs conical quantum dot using the numerical FEM. Indian J. Pure Appl. Phys. **57**, 9 (2019)
31. A. Sali, H. Satori, M. Fliyou, H. Loumrhari, The photoionization cross-section of impurities in quantum dots. Phys. Stat. Sol. (b) **232**, 209–219 (2002)
32. C. Duque, E. Kasapoglu, S. Şakiroglu, H. Sari, I. Sökmen, Intense laser effects on nonlinear optical absorption and optical rectification in single quantum wells under applied electric and magnetic field. Appl. Surf. Sci. **257**, 2313–2319 (2011)
33. G. Rezaei, M. Vahdani, B. Vaseghi, Nonlinear optical properties of a hydrogenic impurity in an ellipsoidal finite potential quantum dot. Curr. Appl. Phys. **11**, 176–181 (2011)
34. H. Dakhlaoui, Tunability of the optical absorption and refractive index changes in step-like and parabolic quantum wells under external electric field. Optik **168**, 416–423 (2018)
35. L.R. Ram-Mohan, S. Saigal, D. Dossa, J. Shertzer, The finite-element method for energy eigenvalues of quantum mechanical systems. Comput. Phys. **4**(1), 50 (1990)
36. H. Satori, A. Sali, The finite element simulation for the shallow impurity in quantum dots. Phys. E **48**, 171–175 (2013)
37. O.C. Zienkiewicz, R.L. Taylor, J.Z. Zhu, *The Finite Element Method: Its Basis and Fundamentals*, 6th edn. (Elsevier, Amsterdam Heidelberg, 2010)
38. A. Sali, H. Satori, The combined effect of pressure and temperature on the impurity binding energy in a cubic quantum dot using the FEM simulation. Superlattices Microstruct. **69**, 38–52 (2014)

MIX
Papier aus verantwortungsvollen Quellen
Paper from responsible sources
FSC® C105338

If you have any concerns about our products,
you can contact us on
ProductSafety@springernature.com

In case Publisher is established outside the EU,
the EU authorized representative is:
Springer Nature Customer Service Center GmbH
Europaplatz 3, 69115 Heidelberg, Germany

Printed by Libri Plureos GmbH
in Hamburg, Germany